T0134913

Studies in Computational Intelligence

Volume 659

About this Series

The series "Studies in Computational Intelligence" (SCI) publishes new developments and advances in the various areas of computational intelligence—quickly and with a high quality. The intent is to cover the theory, applications, and design methods of computational intelligence, as embedded in the fields of engineering, computer science, physics and life sciences, as well as the methodologies behind them. The series contains monographs, lecture notes and edited volumes in computational intelligence spanning the areas of neural networks, connectionist systems, genetic algorithms, evolutionary computation, artificial intelligence, cellular automata, self-organizing systems, soft computing, fuzzy systems, and hybrid intelligent systems. Of particular value to both the contributors and the readership are the short publication timeframe and the worldwide distribution, which enable both wide and rapid dissemination of research output.

More information about this series at http://www.springer.com/series/7092

Jayadeva · Reshma Khemchandani
Suresh Chandra

Twin Support Vector Machines

Models, Extensions and Applications

Jayadeva
Department of Electrical Engineering
Indian Institute of Technology (IIT)
New Delhi
India

Suresh Chandra
Department of Mathematics
Indian Institute of Technology (IIT)
New Delhi
India

Reshma Khemchandani
Department of Computer Science
Faculty of Mathematics and Computer Science
South Asian University (SAU)
New Delhi
India

ISSN 1860-949X ISSN 1860-9503 (electronic)
Studies in Computational Intelligence
ISBN 978-3-319-83462-7 ISBN 978-3-319-46186-1 (eBook)
DOI 10.1007/978-3-319-46186-1

Printed on acid-free paper

This Springer imprint is published by Springer Nature
The registered company is Springer International Publishing AG
The registered company address is: Gewerbestrasse 11, 6330 Cham, Switzerland

To my parents, Shri. R. Sampat Kumaran and Smt. M.C. Ramananda, who have been a lifelong source of inspiration and support.

—Jayadeva

To my loving mother
Mrs. Shilpa Khemchandani, whose beautiful memories will always stay with me,
my family and my teachers.

—Reshma Khemchandani

To my wife Shashi, daughter Surabhi and ever present son Suyash for their continuous support and encouragement.

—Suresh Chandra

Preface

Support vector machines (SVMs), introduced by Vapnik in 1998, have proven very effective computational tool in Machine Learning. SVMs have already outperformed most other computational intelligence methodologies mainly because they are based on sound mathematical principles of statistical learning theory and optimization theory. SVMs have been applied successfully to a wide spectrum of areas, ranging from pattern recognition, text categorization, biomedicine, bioinformatics and brain–computer interface to financial time series forecasting.

In 2007, Jayadeva, Khemchandani and Chandra have proposed a novel classifier called twin support vector machine (TWSVM) for binary data classification. TWSVM generates two non-parallel hyperplanes by solving a pair of smaller-sized quadratic programming problems (QPPs) such that each hyperplane is closer to one class and as far as possible from the other class. The strategy of solving a pair of smaller-sized QPPs, rather than a single large one, makes the learning speed of TWSVM approximately four times faster than the standard SVM.

Over the years, TWSVM has become a popular machine learning tool because of its low computational complexity. Not only TWSVM has been applied to a wide spectrum of areas, many researchers have proposed new variants of TWSVM, for classification, clustering (TWSVC) and regression (TWSVR) scenarios.

This monograph presents a systematic and focused study of the various aspects of TWSVM and related developments for classification, clustering and regression. Apart from presenting most of the basic models of TWSVM, TWSVC and TWSVR available in the literature, a special effort has been made to include important and challenging applications of the tool. A chapter on "Some Additional Topics" has been included to discuss topics of kernel optimization and support tensor machines which are comparatively new, but have great potential in applications.

After presenting an overview of support vector machines in Chap. 1 and devoting an introductory Chapter (Chap. 2) on generalized eigenvalue proximal support vector machines (GEPSVM), the main contents related with TWSVMs are presented in Chaps. 3–8. Here Chap. 8 is fully devoted to "applications".

This monograph is primarily addressed to graduate students and researchers in the area of machine learning and related topics in computer science, mathematics and electrical engineering. Senior undergraduate students having reasonable mathematical maturity will also benefit from this monograph. In writing a monograph of this kind, there is always a difficulty in identifying what to include, or rather what not to include. We have tried our best to include all major variants of the basic TWSVM formulation but still some may have been left out. We hope that the readers will appreciate that a monograph can never be made exhaustive and some such omissions are bound to happen. Although every care has been taken to make the presentation error free, some errors may still remain and we hold ourselves responsible for that and request that errors, if any, be intimated by e-mailing at chandras@maths.iitd.ac.in (e-mail address of S. Chandra).

In the long process of writing this monograph, we have been encouraged and helped by many individuals. We would first and foremost like to thank Prof. Janusz Kacprzyk for accepting our proposal and encouraging us to write this monograph.

We would specially like to thank Prof. O.L. Mangasarian whose research work has been the starting point of our interest in the area of machine learning. In an era when we had no access to the Internet, he has been very kind in sending his reprints/preprints and clarifying many subtle points. In our research, we have been mostly following the work of Prof. Mangasarian and in that sense we have always regarded him as our mentor.

We are highly grateful to Profs. X.J. Peng, Y. Tian and Y.H. Shao and their research collaborators for sending their reprints/preprints and answering to our queries at the earliest. We would also like to thank and acknowledge the editors and the publishers of the journals IEEE Transactions on Pattern Analysis and Machine Intelligence, Neurocomputing, Neural Networks, International Journal of Machine Learning and Cybernetics, Expert System with Applications, Optimization Letters, Information Sciences, European Journal of Operational Research, Neural Computing and its Applications, Applied Intelligence, and Applied Soft Computing for publishing our papers on TWSVM-related work which constitutes the core of this monograph.

Our special thanks are to the authorities of Indian Institute of Technology, Delhi, and South Asian University, New Delhi, for their encouragement and providing all facilities for the preparation of this monograph.

We would like to thank our friends and colleagues Profs. Aparna Mehra, B. Chandra, K.K. Biswas, R.K Sharma, S. Dharmaraja and Shantanu Chaudhary at IIT Delhi, and Danish Lohani, Pranab Muhuri and R.K. Mohanty at South Asian University (SAU) for their interest and encouragement. Professors Amit Bhaya, C.A. Murthy, C.R. Bector, Jeyakumar, Pankaj Gupta, Ravi Kothari, M. Arun Kumar, S. Balasundaram and S.K. Neogy have always encouraged us and took keen interest in the preparation of this monograph.

We appreciate our students Anuj Karpatne, Keshav Goel and Sumit Soman of IIT Delhi, and Pooja Saigal, Aman Pal, Sweta Sharma and Pritam Anand of SAU for their help in the preparation of this manuscript in LATEX and reading the manuscript from a student point of view. Dr. Abha Aggarwal took a major responsibility of organizing the material and her efforts are duly acknowledged.

We are obliged to Dr. Thomas Ditzinger of International Engineering Department, Springer Verlag, for all his help, cooperation and understanding in the publication of this monograph.

New Delhi, India

Jayadeva
Reshma Khemchandani
Suresh Chandra

Contents

Chapter 1
Introduction

1.1 Support Vector Machines: An Overview

Support Vector Machines (SVMs) [1] have emerged to become the paradigm of choice for classification tasks over the last two decades. SVMs emerged from some celebrated work on statistical learning theory. Prior to SVMs, multilayer neural network architectures were applied widely for a variety of learning tasks. Most learning algorithms for multilayer neural networks focussed on minimizing the classification error on training samples. A few efforts focussed on generalization and reducing the error on test samples. Some of the recent works, such as Optimal Brain Surgeon and Optimal Brain Damage, were related to pruning trained networks in order to improve generalization. However, the advent of SVMs radically changed the machine learning landscape. SVMs addressed generalization using a sound theoretical framework and showed that the generalization error was related to the margin of a hyperplane classifier. In order to better appreciate the motivation for Twin Support Vector Machines (TWSVM), we briefly discuss the ideas behind Support Vector Machines. The classical L_1-norm SVM, which minimizes the L_1-norm of the error vector, was followed by formulations that minimize the L_2-norm of the error vector. These includes the Least Squares SVM (LSSVM) proposed by Suykens and Vandewalle [2], and the Proximal SVM (PSVM) proposed by Fung and Mangasarian [3]. The two formulations are similar in spirit, but have minor differences. The classical L_1-norm SVM and its least squares counterparts suffer from one drawback. The formulations work well for binary classification when the two classes are balanced, but for imbalanced classes, nonlinear separable datasets, test set accuracies tend to be poor. One can construct synthetic datasets to reveal that in this scenario, the classifying hyperplane tends to increasingly ignore the smaller class, and that test set accuracies drop for the smaller class with increasing imbalance in the class sizes.

The Generalized Eigenvalue Proximal SVM (GEPSVM) is an attempt to include information within samples of the same class and between samples of different classes. It represents a radically different approach to the classical SVM, and is less

Jayadeva et al., *Twin Support Vector Machines*, Studies in Computational Intelligence 659, DOI 10.1007/978-3-319-46186-1_1

sensitive to class size imbalance. The GEPSVM leads to solving eigenvalue problem, and no longer requires solving a Quadratic Programming Problem as in the case of the L_1-norm and L_2-norm SVMs. The GEPSVM is an important milestone in the journey from SVMs to the TWSVM, and is discussed in more detail in Chap. 2.

This chapter consists of the following main sections: The Classical L_1-norm SVM, Least Squares SVM and Proximal SVM, Support Vector Regression (SVR), Efficient Algorithms for SVM and SVR, and other approaches for solving the quadratic programming problems arising in the SVM type formulations. The presentation of this chapter is based on the work of Bi and Bennett [4], Suykens and Vandewalle [2], Fung and Mangasarian [3], Burges [5], Sastry [6], Deng et al. [7].

1.2 The Classical L_1-Norm SVM

Consider a binary classification dataset in which the training samples are denoted by

$$T_C = \{(x^{(i)}, y_i), \quad (i = 1, 2, \ldots, m)\}, \tag{1.1}$$

where $x^{(i)} \in \mathbb{R}^n$ and $y_i \in \{-1, 1\}, i = 1, 2, \ldots, m$. Let there be m_1 patterns having class label $+1$ and m_2 patterns having class label -1. We construct matrix A (respectively B) of order $(m_1 \times n)$ (respectively $(m_2 \times n)$) by taking the ith row of A (respectively B) as the ith pattern of class label $+1$ (respectively class label -1). Thus, $m_1 + m_2 = m$.

1.2.1 Linear SVM: Hard Margin Classifier

The linear L_1-norm SVM attempts to find a hyperplane of the form

$$w^T x + b = 0, \tag{1.2}$$

with maximum margin that separates samples of the two classes. Let us initially consider the case when the samples of A and B are linearly separable, i.e., samples of A lie on one side of the separating hyperplane, and samples of B lie on the other side. To determine the classifier $w^T x + b = 0$ for the linearly separable dataset T_c, Vapnik [1] proposed the principle of maximum margin. According to this principle, we should determine $w \in \mathbb{R}^n$, $b \in \mathbb{R}$ so that the margin, i.e. the distance between the supporting hyperplanes $w^T x + b = 1$ and $w^T x + b = -1$ is maximum. A directed calculation yields that the margin equals $\frac{2}{||w||}$, where $||w||$ denotes the L_2-norm of $w \in \mathbb{R}^n$ and is given by $||w||^2 = w^T w$.

Though the principle of maximum margin is derived through certain inequalities/bounds involving the concept of Vapnik–Chervonenkis (VC) dimension, one can justify it on the grounds that the larger is the margin, the smaller is the probability that a hyperplane will determine the class of a test sample incorrectly. Therefore, the maximum margin SVM is obtained by solving the following optimization problem.

$$\underset{(w,b)}{\text{Max}} \quad \frac{2}{\|w\|}$$
subject to

$$\begin{aligned} Aw + eb &\geq 1, \\ Bw + eb &\leq -1, \end{aligned} \tag{1.3}$$

where e is the vector of 'ones' of appropriate dimension.

Since maximizing $\frac{2}{\|w\|}$ is equivalent to minimizing $\frac{1}{2}\|w\|$, the above problem may be re-written as

$$\underset{(w,b)}{\text{Min}} \quad \frac{1}{2}\|w\|$$
subject to

$$\begin{aligned} Aw + eb &\geq 1, \\ Bw + eb &\leq -1. \end{aligned} \tag{1.4}$$

Since $\|w\|$, the norm of the weight vector, is non-negative, we can also replace the objective function by its square, yielding the problem

$$\underset{(w,b)}{\text{Min}} \quad \frac{1}{2}\|w\|^2$$
subject to

$$\begin{aligned} Aw + eb &\geq 1, \\ Bw + eb &\leq -1. \end{aligned} \tag{1.5}$$

This is a convex quadratic programming problem (QPP) and is termed as the Hard-Margin Linear SVM. The formulation may be more succinctly written as

$$\underset{(w,b)}{\text{Min}} \quad \frac{1}{2}w^T w$$
subject to

$$y_i(w^T x^{(i)} + b) \geq 1, \quad (i = 1, 2, \ldots, m). \tag{1.6}$$

Here it may be noted that problem (1.6) has as many constraints as the number of training samples, i.e., m. Since in practice m could be very large, it makes sense to write the Wolfe dual [8, 9] of problem (1.6), which comes out to be

$$\begin{aligned}
\underset{\alpha}{\text{Max}} \quad & -\frac{1}{2}\sum_{i=1}^{m}\sum_{j=1}^{m} y_i y_j (x^{(i)})^T x^{(j)} \alpha_i \alpha_j + \sum_{j=1}^{m} \alpha_j \\
\text{subject to} \quad & \\
& \sum_{i=1}^{m} y_i \alpha_i = 0, \\
& \alpha_i \geq 0, \quad (i = 1, 2, \ldots, m).
\end{aligned} \tag{1.7}$$

Let $\alpha^\star = (\alpha_1^\star, \alpha_2^\star, \ldots, \alpha_m^\star)$ be an optimal solution of (1.7). Then using Karush–Kuhn–Tucker (K.K.T) [9] conditions it can be shown that the Hard Margin classifier is $x^T w^\star + b^\star = 0$ where

$$w^\star = \sum_{i=1}^{m} \alpha_i^\star y_i x^{(i)}, \tag{1.8}$$

$$b^\star = y_j - \sum_{i=1}^{m} \alpha_i^\star y_i (x^{(i)})^T x^{(j)}. \tag{1.9}$$

Here, $x^{(j)}$ is that pattern for which $\alpha_j^\star > 0$. All patterns $x^{(j)}$ for which $\alpha_j^\star > 0$ are called as support vectors. In view of K.K.T conditions for problem (1.6), we get $y_j((w^\star)^T x^{(j)} + b) = 1$, i.e., $x^{(j)}$ lies on one of the bounding hyperplanes, $w^T x + b = 1$ and $w^T x + b = -1$. Since there can be many support vectors, in practice $b^\star$ is taken as the average of all such $b^\star$.

1.2.2 Linear SVM: Soft Margin Classifier

Consider a case where a few samples are not linearly separable. In this case, one may choose to allow some training samples to be mis-classified, and to seek a classifying hyperplane that achieves a tradeoff between having a large margin and a small classification error. The optimization problem that allows us to find such a tradeoff is given by

$$\begin{aligned}
\underset{(w,b,\xi)}{\text{Min}} \quad & \frac{1}{2} w^T w + C \sum_{i=1}^{m} \xi_i \\
\text{subject to} \quad & \\
& y_i (w^T x^{(i)} + b) + \xi_i \geq 1, \quad (i = 1, 2, \ldots, m), \\
& \xi_i \geq 0, \quad (i = 1, 2, \ldots, m),
\end{aligned} \tag{1.10}$$

and yields the soft-margin SVM. The above formulation is the primal formulation and is usually not the one that is solved.

The margin is termed as *soft* because the hyperplane $w^T x + b = 0$ does not linearly separate samples of the two classes, and some samples are incorrectly classified. The term $\sum_{i=1}^{m} \xi_i$ measures the mis-classification error, and the hyper-parameter C defines the importance of the mis-classification term $\sum_{i=1}^{m} \xi_i$ relative to the margin term $\frac{1}{2} w^T w$. This hyper-parameter is chosen by the user and is often determined by using a tuning set, that is a small sample of the training set, on which a search is conducted to find a good choice of C. Once again, we can determine the Wolfe's dual of the primal soft margin. A little algebra yields the dual as

$$\begin{aligned} \underset{\alpha}{\text{Max}} \quad & -\frac{1}{2} \sum_{i=1}^{m} \sum_{j=1}^{m} y_i y_j (x^{(i)})^T x^{(j)} \alpha_i \alpha_j + \sum_{j=1}^{m} \alpha_j \\ \text{subject to} \quad & \\ & \sum_{i=1}^{m} y_i \alpha_i = 0, \\ & 0 \le \alpha_i \le C, \quad (i = 1, 2, \ldots, m). \end{aligned} \tag{1.11}$$

Let $\alpha^\star = (\alpha_1^\star, \alpha_2^\star, \ldots, \alpha_m^\star)$ be an optimal solution of (1.11). Then using K.K.T conditions, it can be shown that the soft margin classifier is $x^T w^\star + b^\star = 0$ where

$$w^\star = \sum_{i=1}^{m} \alpha_i^\star y_i x^{(i)}. \tag{1.12}$$

In this case, the support vectors are those patterns $x^{(j)}$ for which the corresponding multiplier α_j satisfies $0 < \alpha_j < C$; using the K.K.T conditions, one can also show that

$$y_j (w^{\star T} x^{(j)} + b) = 1, \tag{1.13}$$

i.e., the support vectors are samples that lie on the hyperplanes

$$w^{\star T} x^{(j)} + b^\star = \pm 1.$$

In the case of the soft margin SVM, the offset $b^\star$ is obtained by taking any support vector, i.e. any sample $x^{(j)}$ for which $0 < \alpha_j < C$; and computing,

$$b^\star = y_j - \sum_{i=1}^{m} \alpha_i^\star y_i (x^{(i)})^T x^{(j)}. \tag{1.14}$$

Since there can be many support vectors, in practice $b^\star$ is taken as the average of all such $b^\star$ obtained from all the support vectors.

One may also note that samples $x^{(j)}$ for which $\alpha_j = 0$ are not support vectors and do not effect the classifying hyperplane. Finally, samples $x^{(j)}$ for which $\alpha_j = C$ are samples that are classified with a margin less than 1, i.e.

$$y_j(w^{\star T} x^{(j)} + b^\star) < 1,$$

are also support vectors, since their corresponding multipliers are non-zero. These also include samples that are mis-classified.

When the data sets are too complex to be separated adequately well by a linear hyperplane, the solution is to employ the so-called "kernel trick". This involves projecting the input samples to a higher dimensional "image" space or "feature" space, and constructing a linear hyperplane in that space. This leads to the nonlinear or kernel SVM, which we discuss in the next section.

1.2.3 Nonlinear/Kernel SVM

Consider the exclusive-OR (XOR) problem in two dimensions. Here, the samples for the binary classification problem are $(x_1, x_2) = (0, 0), (1, 1), (0, 1)$ and (1,0). The first two samples are in class (-1), while the latter two are in class 1, i.e., the class labels are $\{-1, -1, 1, 1\}$. The samples are not linearly separable and a linear classifier of the form $w_1 x_1 + w_2 x_2 + b = 0$ will be unable to separate the patterns. Consider a map $\phi(x)$ that maps the input patterns to a higher dimension z. As an example, we choose a map that takes the patterns (x_1, x_2) and maps them to three dimensions, i.e. $z = \phi(x)$, where $z = (z_1, z_2, z_3)$, with

$$\begin{aligned} z_1 &= x_1, \\ z_2 &= x_2, \\ z_3 &= -(2x_1 - 1) \times (2x_2 - 1). \end{aligned} \tag{1.15}$$

The map ϕ thus produces the following patterns in the image space:

$$\begin{aligned} (0, 0) \xrightarrow{\phi} (0, 0, -1), \text{ class label } &= -1, \\ (1, 1) \xrightarrow{\phi} (0, 0, -1), \text{ class label } &= -1, \\ (0, 1) \xrightarrow{\phi} (0, 1, +1), \text{ class label } &= +1, \\ (1, 0) \xrightarrow{\phi} (1, 0, +1), \text{ class label } &= +1. \end{aligned}$$

Note that the plane $z_3 = 0$ separates the patterns. The map ϕ from two dimensions to three has made the patterns separable in the higher (three) dimensional feature space.

In order to construct the SVM classifier, we consider the patterns $\phi(x^{(i)}$, $(i = 1, 2, \ldots, m)$, with corresponding class labels y_i, $(i = 1, 2, \ldots, m)$, and rewrite the soft margin SVM classifier given by (1.11) for the given patterns. The soft margin kernel SVM is given by

$$\begin{aligned}
&\underset{(w,b,\xi)}{\text{Min}} \quad \frac{1}{2}w^T w + C\sum_{i=1}^{m}\xi_i \\
&\text{subject to} \\
&\qquad y_i(w^T\phi(x^{(i)}) + b) + \xi_i \geq 1, \quad (i = 1, 2, \ldots, m), \\
&\qquad \xi_i \geq 0, \quad (i = 1, 2, \ldots, m).
\end{aligned} \tag{1.16}$$

The corresponding dual is similarly obtained by substituting $\phi(x^{(i)})$ for $x^{(i)}$ in (1.11). This yields the Wolfe dual of the soft margin kernel SVM, as

$$\begin{aligned}
&\underset{\alpha}{\text{Max}} \quad -\frac{1}{2}\sum_{i=1}^{m}\sum_{j=1}^{m} y_i y_j \phi(x^{(i)})^T\phi(x^{(j)})\alpha_i\alpha_j + \sum_{j=1}^{m}\alpha_j \\
&\text{subject to} \\
&\qquad \sum_{i=1}^{m} y_i\alpha_i = 0, \\
&\qquad 0 \leq \alpha_i \leq C, \quad (i = 1, 2, \ldots, m).
\end{aligned} \tag{1.17}$$

Note that the image vectors $\phi(x^{(i)})$ are only present as part of inner product terms of the form $\phi(x^{(i)})^T\phi(x^{(j)})$. Let us define a *Kernel function* K as follows,

$$K(p, q) = \phi(p)^T\phi(q). \tag{1.18}$$

Observe that the value of a kernel function is a scalar obtained by computing the inner product of the image vectors $\phi(p)$ and $\phi(q)$. It is possible to obtain that value by implicitly using the values of $\phi(p)$ and $\phi(q)$, and not explicitly. The implication is that one can take two vectors p and q that lie in the input space, and obtain the values of $\phi(p)^T\phi(q)$ without explicitly computing $\phi(p)^T$ or $\phi(q)$. Since the map ϕ takes input vectors from a lower dimensional input space and produces high dimensional image vectors, this means that we do not need to use the high dimensional image vectors at all. This requires the existence of a Kernel function K so that $K(p, q) = \phi(p)^T\phi(q)$

for any choice of vectors p and q. A theorem by Mercer [6] provides the necessary conditions for such a function to exist.

Mercer's Theorem

Consider a kernel function $K : \mathbb{R}^n \times \mathbb{R}^n \to \mathbb{R}$. Then, for such a kernel function, there exists a map ϕ such that $K(p, q) \equiv \phi(p)^T \phi(q)$ for any vectors p and q if and only if the following condition holds.

For every square integrable function $h : \mathbb{R}^n \to \mathbb{R}$, i.e. $\int h^2(x)dx < \infty$, we have

$$\int K(p, q)h(p)h(q)dpdq \geq 0.$$

It is easy to show that if $K(p, q)$ is expandable as a convergent series with finite or infinite polynomial terms with non-negative co-efficients, i.e. it is of the form $K(p, q) = \sum_{j=0}^{\infty} \alpha_j (p^T q)^j$ with $\alpha_j \geq 0$, then $K(p, q)$ satisfies the conditions of Mercer's theorem.

Kernels commonly employed with SVMs include: The linear kernel, with

$$K(p, q) = p^T q,$$

and the polynomial kernel with degree d, i.e.

$$K(p, q) = (a + p^T q)^d = a^d + da^{d-1}(p^T q) + \cdots + (p^T q)^d.$$

Note that the expansion contains only non-negative terms. A very popular kernel is the Radial Basis Function (RBF) or Gaussian kernel, i.e.

$$K(p, q) = exp(-\beta \|p - q\|^2), \tag{1.19}$$

which may also be shown to have an infinite expansion of non-negative terms.

The dual formulation is now better expressed in terms of elements of the Kernel matrix $K = [K_{ij}]_{m \times m}$, where $K_{ij} = K(x^{(i)}, x^{(j)})$, and is given by

$$\underset{\alpha}{\text{Max}} \quad -\frac{1}{2} \sum_{i=1}^{m} \sum_{j=1}^{m} y_i y_j K_{ij} \alpha_i \alpha_j + \sum_{j=1}^{m} \alpha_j$$

subject to

$$\sum_{i=1}^{m} y_i \alpha_i = 0,$$
$$0 \leq \alpha_i \leq C, \quad (i = 1, 2, \ldots, m). \tag{1.20}$$

1.3 Least Squares SVM and Proximal SVM

Many of the early SVM formulations largely differ with regard to the way they measure the empirical error. When the L_2-norm, i.e. the squared or Euclidean norm is used to measure the classification error, we obtain the Least Squares SVM (LSSVM) proposed by Suykens and Vandewalle [2], and the Proximal SVM (PSVM) proposed by Fung and Mangasarian [3]. The two differ in minor ways in terms of their formulations but employ similar methodology for their solutions. We shall write LSSVM (respectively PSVM) for least square SVM (respectively Proximal SVM).

The LSSVM solves the following optimization problem.

$$\underset{(w,b,\xi)}{\text{Min}} \quad \frac{1}{2}\|w\|^2 + C\sum_{i=1}^{m}(\xi_i)^2$$

subject to

$$y_i(w^T x^{(i)} + b) = 1 - \xi_i, \quad (i = 1, 2, \ldots, m), \tag{1.21}$$

while the PSVM solves the problem

$$\underset{(w,b,\xi)}{\text{Min}} \quad \frac{1}{2}(w^T w + b^2) + C\sum_{i=1}^{m}(\xi_i)^2$$

subject to

$$y_i(w^T x^{(i)} + b) = 1 - \xi_i, \quad (i = 1, 2, \ldots, m). \tag{1.22}$$

The addition of the term b^2 to the objective function makes it strictly convex (Fung and Mangasarian [3]), and facilitates algebraic simplification of the solution. The evolution of the GEPSVM and the TWSVM are based on a similar premise, and in the sequel we focus on the PSVM and its solution. Note that in the PSVM formulation the error variables ξ_i are unrestricted in sign.

Similar to LSSVM, the solution of PSVM aims at finding $w^T x + b = \pm 1$ which also represents a pair of parallel hyperplanes. Note that $w^T x + b = 1$ is the representation of all the samples of class A, while $w^T x + b = -1$ is the representation of all the samples of class B respectively. The hyperplane $w^T x + b = 1$ can thus be thought of as a hyperplane that passes through the cluster of samples of A. It is therefore termed as a hyperplane that is "proximal" to the set of samples of A. In a similar manner, the hyperplane $w^T x + b = -1$ can thus be thought of as a hyperplane that passes through the cluster of samples of B. The first term of the objective function of (1.22) is the reciprocal of the distance between the two proximal hyperplanes, and minimizing it attempts to find two planes that pass through samples of the two classes, while being as far apart as possible. The TWSVM is motivated by a similar notion, as we shall elaborate in the sequel, but it offers a more generalized solution by allowing the two hyperplanes to be non-parallel.

The formulation (1.22) is also a regularized least squares solution to the system of linear equations

$$A^i w + b = 1, \; (i = 1, 2, \ldots, m_1), \tag{1.23}$$

$$B^i w + b = -1, \; (i = 1, 2, \ldots, m_2). \tag{1.24}$$

The Lagrangian for the system (1.22) in a vector form is given by

$$\begin{aligned} L(w, b, \xi^A, \xi^B, \lambda^A, \lambda^B) = {} & C(\|\xi^A\|^2 + \|\xi^B\|^2) - (\lambda^A)^T(Aw + e^A b + \xi^A - e^A) \\ & + \frac{1}{2}(w^T w + b^2) - (\lambda^B)^T(Bw + e^B b + \xi^B + e^B), \end{aligned} \tag{1.25}$$

where ξ_A and ξ_B are error variables; e^A and e^B are vectors of ones of appropriate dimension. Here, λ^A and λ^B are vectors of Lagrange multipliers corresponding to the equality constraints associated with samples of classes A and B, respectively. Setting the gradients of L to zero yields the following Karush–Kuhn–Tucker optimality conditions

$$w - A^T\lambda^A - B^T\lambda^B = 0 \implies w = A^T\lambda^A + B^T\lambda^B, \tag{1.26}$$

$$b - (e^A)^T\lambda^A - (e^B)^T\lambda^B = 0 \implies b = (e^A)^T\lambda^A + (e^B)^T\lambda^B, \tag{1.27}$$

$$C\xi^A - \lambda^A = 0 \implies \xi^A = \frac{\lambda^A}{C}, \tag{1.28}$$

$$C\xi^B - \lambda^B = 0 \implies \xi^B = \frac{\lambda^B}{C}, \tag{1.29}$$

$$Aw + e^A b + \xi^A = e^A, \tag{1.30}$$

$$Bw + e^B b + \xi^B = -e^B. \tag{1.31}$$

We do not proceed with the algebraic simplification of the conditions to obtain the PSVM solution. The interested reader is referred to the original work by Fung and Mangasarian [3] for details.

Figure 1.1 shows an example with synthetic data samples in two dimensions. The example is peculiar because one of the classes has an outliers that is far from the cluster of samples of its class. The SVM solution is shown in Fig. 1.2 and the PSVM planes are shown in Fig. 1.3. This example illustrates that the SVM can be more sensitive to outliers in the data.

One assumption made in the SVM and the PSVM, though implicitly, is that the distributions of the two classes A and B are similar. This is usually not the case. Hence, parallel planes may not be able to separate classes well when one of them has a much larger variance than the other. This motivates us to look for formulations that

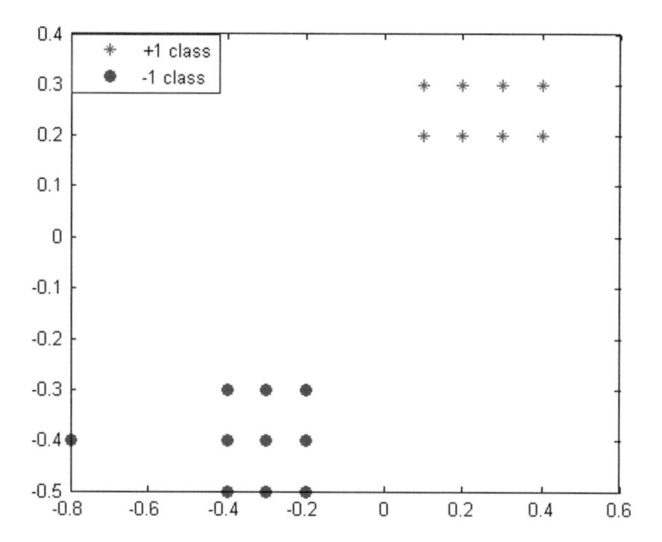

Fig. 1.1 Synthetic dataset with an outlier for one class

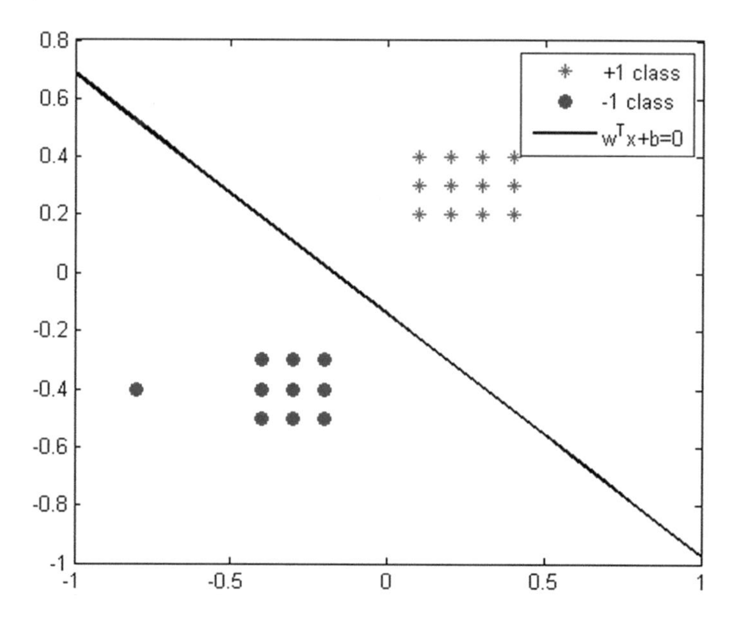

Fig. 1.2 SVM solution for the dataset

can take into account the scatter within each class, and the between class scatter. This idea may be thought of as a motivation for the Generalized Eigen Value Proximal Support Vector Machine (GEPSVM). The basic GEPSVM model of Mangasarian and Wild [10] is a major motivation for ideas behind Twin Support Vector Machines.

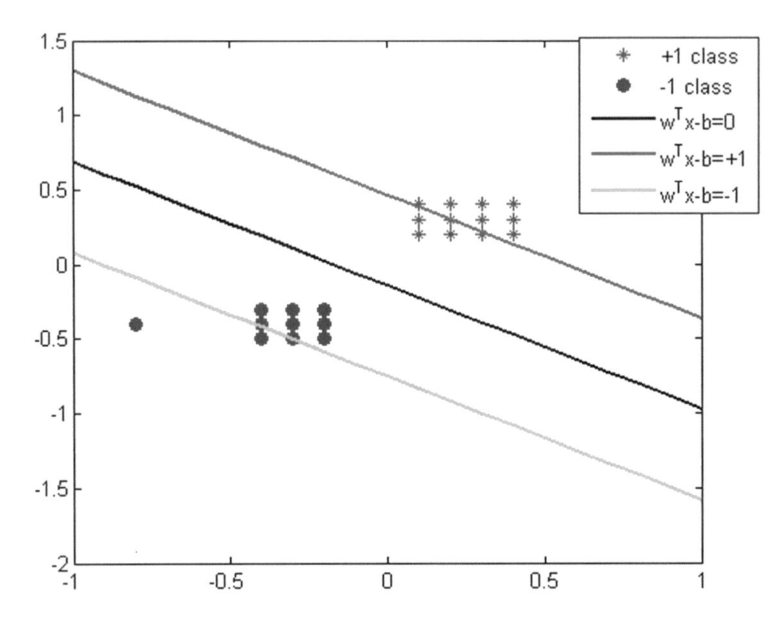

Fig. 1.3 PSVM solution for the dataset

1.4 Support Vector Regression

The task of regression consists of determining a regressor $f(x)$ so that $f(x^{(i)}) \approx y_i$, where $y_i \in \mathbb{R}$ is the value of a measurement made at the sample location $x^{(i)}$. Classification may be thought of as a special case of regression, in which the values of $y_i \in \{-1, 1\}$. Linear regression attempts to find a regressor of the form

$$f(x) = w^T x + b.$$

We examine the case for ϵ-regression, where the task is to find a regressor that approximately matches the samples y_i within a tolerance ϵ.

In analogy with the hard margin SVM classifier, the hard ϵ-tube regression problem is usually formulated as

$$\underset{(w,b)}{\text{Min}} \quad \frac{1}{2}\|w\|^2$$

subject to

$$\begin{aligned} w^T x^{(i)} + b &\geq y_i - \epsilon, \quad (i = 1, 2, \ldots, m), \\ w^T x^{(i)} + b &\leq y_i + \epsilon, \quad (i = 1, 2, \ldots, m). \end{aligned} \tag{1.32}$$

But this formulation does not have a correspondence with the SVM classification problem, and hence, the notion of margin cannot be adapted to the regression setting. This requires us to use a clever and significant result due to Bi and Bennett [4], that shows a link between the classification and regression tasks. Consider a regression problem with data points $x^{(i)}$, $(i = 1, 2, \ldots, m)$, and where the value of an unknown function at the point $x^{(i)}$ is denoted by $y_i \in \mathbb{R}$. As before, we assume that the dimension of the input samples is n, i.e. $x^{(i)} = (x_1^{(i)}, x_2^{(i)}, \ldots, x_n^{(i)})^T$. Bi and Bennett [4] showed that the task of building a regressor on this data has a one-to-one correspondence with a binary classification task in which class (-1) points lie at the $(n+1)$-dimensional co-ordinates $(x^{(1)};\ y_1 - \epsilon), (x^{(2)};\ y_2 - \epsilon), \ldots, (x^{(m)};\ y_m - \epsilon)$, and class $(+1)$ points lie at the co-ordinates $(x^{(1)}; y_1 + \epsilon),\ (x^{(2)}; y_2 + \epsilon),\ \ldots,\ (x^{(m)}; y_m + \epsilon)$. Let us first consider the case when these two subsets are linearly separable. This implies that the convex hulls of these two subsets do not intersect. Let the closest points in the two convex hulls be denoted by $p+$ and $p-$, respectively. The maximum margin separating hyperplane must pass through the midpoint of the line joining $p+$ and $p-$. Let the separating hyperplane be given by $w^T x + \eta y + b = 0$. We observe that $\eta \neq 0$. Then, the regressor is given by

$$y = -\frac{1}{\eta}(w^T x + b). \tag{1.33}$$

Thus, as shown in Bi and Bennett [4] the SVR regression problem is written as

$$\begin{aligned}
&\underset{(w,\eta,\alpha,\beta)}{\text{Min}} \quad \frac{1}{2}(\|w\|^2 + \eta^2) + (\beta - \alpha) \\
&\text{subject to}
\end{aligned}$$

$$\begin{aligned}
&w^T x^{(i)} + \eta(y_i + \epsilon) \geq \alpha, \quad (i = 1, 2, \ldots, m), \\
&w^T x^{(i)} + \eta(y_i - \epsilon) \leq \beta, \quad (i = 1, 2, \ldots, m).
\end{aligned} \tag{1.34}$$

Remark 1.4.1 By taking $\alpha = (1 - b)$ and $\beta = -(1 + b)$, we get $\alpha - \beta = 2$ and in that case the hard ϵ-tube formulation (1.34) reduces to the problem

$$\begin{aligned}
&\underset{(w,b,\eta)}{\text{Min}} \quad \frac{1}{2}(\|w\|^2 + \eta^2) \\
&\text{subject to}
\end{aligned}$$

$$\begin{aligned}
&w^T x^{(i)} + \eta(y_i + \epsilon) + b \geq 1, \quad (i = 1, 2, \ldots, m), \\
&w^T x^{(i)} + \eta(y_i - \epsilon) + b \leq -1, \quad (i = 1, 2, \ldots, m).
\end{aligned} \tag{1.35}$$

Here it may be noted that problem (1.35) is same as the problem obtained in Deng et al. [7], which has been reduced to the usual ϵ-SVM formulation (1.32). Some of these details have been summarized in Chap. 4.

We can write the Wolfe dual of (1.34) and get the following QPP

$$\begin{aligned} \underset{(u,v)}{\text{Min}} \quad & \frac{1}{2}\left\| \begin{pmatrix} X^T \\ (y+\epsilon e)^T \end{pmatrix} u - \begin{pmatrix} X^T \\ (y-\epsilon e)^T \end{pmatrix} v \right\|^2 \\ \text{subject to} \quad & \\ & e^T u = 1, \; e^T v = 1, \\ & u \geq 0, \; v \geq 0, \end{aligned} \tag{1.36}$$

where $X = (x^{(1)}, x^{(2)}, \ldots, x^{(m)})^T \in \mathbb{R}^{m \times n}$, $y \in \mathbb{R}^m$, $u \in \mathbb{R}^{m+1}$ and $v \in \mathbb{R}^{m+1}$.

Let $(\widehat{u}, \widehat{v})$ be optimal to (1.36). Then $\widehat{w} = X^T(\widehat{u} - \widehat{v})$, $\widehat{\eta} = (y + \epsilon e)^T \widehat{u} - (y - \epsilon e)^T \widehat{v}$ and $-b = \frac{1}{2}(\widehat{W})^T (X^T(\widehat{u} + \widehat{v}) + \widehat{\eta}\, y^T(\widehat{u} + \widehat{v}))$. Then the separating plane has the equation $(\widehat{W})^T X + \widehat{\eta}\, y + \widehat{b} = 0$. Rescaling this plane by $-\widehat{\eta}$ yields the regression function $y = -\frac{1}{\widehat{\eta}}((\widehat{w})^T x + b)$.

The soft ϵ-tube version of the ϵ-SVR may be written as

$$\begin{aligned} \underset{(w,\xi^+,\xi^-,\eta,\alpha,\beta)}{\text{Min}} \quad & \frac{1}{2}(\|w\|^2 + \eta^2) + (\beta - \alpha) + C\sum_{i=1}^{m}(\xi_i^+ + \xi_i^-) \\ \text{subject to} \quad & \\ & w^T x^{(i)} + \xi_i^+ + \eta(y_i + \epsilon) \geq \alpha, \\ & w^T x^{(i)} - \xi_i^- + \eta(y_i - \epsilon) \leq \beta, \\ & \xi_i^+, \xi_i^- \geq 0, \quad (i = 1, 2, \ldots m). \end{aligned} \tag{1.37}$$

We can write the Wolfe dual of (1.37) to get

$$\begin{aligned} \underset{(u,v)}{\text{Min}} \quad & \frac{1}{2}\| \begin{pmatrix} X^T \\ (y+\epsilon e)^T \end{pmatrix} u - \begin{pmatrix} X^T \\ (y-\epsilon e)^T \end{pmatrix} v \|^2 \\ \text{subject to} \quad & \\ & e^T u = 1, \; e^T v = 1, \\ & 0 \leq u \leq Ce, \; 0 \leq v \leq Ce. \end{aligned} \tag{1.38}$$

Let $(\widehat{u}, \widehat{v})$ be optimal to (1.38). Then $\widehat{w} = X^T(\widehat{u} - \widehat{v})$, $\widehat{\eta} = (y + \epsilon e)^T \widehat{u} - (y - \epsilon e)^T \widehat{v}$ and $-b = \frac{1}{2}(\widehat{W})^T (X^T(\widehat{u} + \widehat{v}) + \widehat{\eta}\; y^T(\widehat{u} + \widehat{v}))$. Then the required regressor is $y = -\frac{1}{\widehat{\eta}}(\widehat{w}^T x + \widehat{b})$.

The kernel version of the ϵ-SVR is obtained on substituting $\phi(X)$ for X and following the standard methodology of SVM.

Remark 1.4.2 Bi and Bennett [4] have termed the hard ϵ-tube formulation as H-SVR (convex hull SVR) because it is constructed by separating the convex hulls of the training data with the response variable shifted up and down by ϵ. The soft ϵ-tube formulation, termed as RH-SVR, is constructed by separating the reduced convex hulls of the training data with the response variable shifted up and down by ϵ. Reduced convex hulls limit the influence of any given point by reducing the upper bound on the multiplier for each point to a constant $D < 1$. There is a close connection between the constant D (appearing in RH-SVR) and C (appearing in soft ϵ-tube SVR). For these and other details we shall refer to Bi and Bennett [4].

Remark 1.4.3 An alternate but somewhat simpler version of the derivation of ϵ-SVR is presented in Deng et al. [7] which is also summarized in Chap. 4. The presentation in Deng et al. [7] is again based on Bi and Bennett [4].

1.5 Efficient Algorithms for SVM and SVR

So far, we have seen formulations for Support Vector Machines for classification and for regression. All the formulations involve solving a quadratic programming problem (QPP), i.e., minimizing a quadratic objective function with linear constraints. While there are several algorithms for solving QPPs, the structure of the QPP in SVMs allows for the construction of particularly efficient algorithms. One of the early algorithms for doing so was given by Platt [11], and is referred to as Platt's Sequential Minimal Optimization (SMO) technique.

We first begin by re-iterating the dual formulation of the SVM, for the sake of convenience. The dual QPP formulation for the soft margin SVM classifier is given by

$$\underset{\alpha}{\text{Max}} \quad -\frac{1}{2}\sum_{i=1}^{m}\sum_{j=1}^{m} y_i y_j K_{ij}\alpha_i\alpha_j + \sum_{j=1}^{m}\alpha_j$$

subject to

$$\sum_{i=1}^{m} y_i\alpha_i = 0,$$
$$0 \le \alpha_i \le C, \quad (i = 1, 2, \ldots, m). \tag{1.39}$$

Platt's SMO starts with all multipliers α_i, $(i = 1, 2, \ldots, m)$ set to zero. This trivially satisfies the constraints of (1.39). The algorithm updates two multipliers at a time. It first selects one multipliers to update, and updates it in order to achieve the maximum reduction in the value of the objective function. Note that changing the value of any multiplier would now lead to constraints of (1.39) getting violated. Platt's

SMO algorithm now updates the second multiplier to ensure that the constraints of (1.39) remain satisfied. Thus, each iteration ensures a reduction in the value of the objective function without violating any of the constraints. We now elaborate on the individual steps of the algorithm.

Step 1

The first step is to find a pair of violating multipliers. This involves checking each multiplier α_i, $(i = 1, 2, \ldots, m)$ to see if it violates the K.K.T conditions. These conditions may be summarized as:

(i) $\alpha_i = 0 \implies y_i f(x^{(i)}) \geq 1$,

(ii) $0 < \alpha_i < C \implies y_i f(x^{(i)}) = 1$,

(iii) $\alpha_i = C \implies y_i f(x^{(i)}) < 1$,

where

$$f(p) = \sum_{j=1}^{m} \lambda_j y_j K(p, x^{(j)}) + b. \tag{1.40}$$

If there are no violating multipliers, then the existing solution is optimal and the algorithm terminates. Otherwise, it proceeds to Step 2. For the sake of convenience, we assume that the violators chosen in Step 1 are denoted by α_1 and α_2. Although one could pick any pair of violating multipliers in Step 1, picking the most violating pairs leads to faster convergence and fewer iterations. More sophisticated implementations of SMO indeed do that in order to reduce the number of updates needed for convergence.

Step 2

Since only α_1 and α_2 are changed, and since $\sum_{i=1}^{m} y_i \alpha_i = 0$, this implies that the updated values of α_1 and α_2, denoted by α_1^{next} and α_2^{next} must satisfy

$$\alpha_1^{next} y_1 + \alpha_2^{next} y_2 = \alpha_1 y_1 + \alpha_2 y_2 = \beta \text{ (say)}, \tag{1.41}$$

where α_1 and α_2 indicate their current values. Hence, we have

$$\alpha_1^{next} = \beta - y_1 y_2 \alpha_2^{next}, \tag{1.42}$$

since $y_1, y_2 \in \{-1, 1\}$. Note that the updated values α_1^{next} and α_2^{next} need to satisfy $0 \leq \alpha_1^{next}, \alpha_2^{next} \leq C$. The constraints on α_1^{next} and α_2^{next} may be summarized as $\alpha_{min} \leq \alpha_2^{next} \leq \alpha_{max}$, together with (1.42), where α_{min} and α_{max} are indicated in the

Table 1.1 α_{min} and α_{max}

	$y_1 \neq y_2$	$y_1 = y_2$
α_{min}	$max(0, \alpha_2 - \alpha_1)$	$max(0, \alpha_2 + \alpha 1 - C)$
α_{max}	$min(C, \alpha_2 - \alpha_1 + C)$	$min(C, \alpha_2 + \alpha_1)$
β	$\alpha_1 - \alpha_2$	$(\alpha_1 + \alpha_2)$

following table. The Table 1.1 also indicates the value of β in terms of the current values of α_1 and α_2.

Since only α_1 and α_2 are being changed, the objective function of the SVM formulation, as given in (1.39) may be written as

$$Q(\alpha_1, \alpha_2) = -\tfrac{1}{2}(K_{11}\alpha_1^2 + K_{22}\alpha_2^2 + 2K_{12}y_1y_2\alpha_1\alpha_2 + \alpha_1 y_1 z_1 + \alpha_2 y_2 z_2) \\ +(\alpha_1 + \alpha_2) + R(\alpha \setminus \{\alpha_1, \alpha_2\}), \tag{1.43}$$

where $R(\alpha \setminus \{\alpha_1, \alpha_2\})$ is a term that depends on $\alpha_i, i \neq 1, 2$, and

$$z_1 = \sum_{j=3}^{M} y_j \alpha_j K_{1j}, \tag{1.44}$$

$$z_2 = \sum_{j=3}^{M} y_j \alpha_j K_{2j}. \tag{1.45}$$

Since $\alpha_1^{next} = \beta - y_1 y_2 \alpha_2^{next}$ from (1.42), we can further simplify $Q(\alpha_1, \alpha_2)$ as a function of α_2 alone, i.e.

$$Q(\alpha_2) = -\frac{1}{2}(K_{11}(\beta - \alpha_2)^2 + K_{22}\alpha_2^2 + 2K_{12}y_1y_2\alpha_2(\beta - \alpha_2) + (\beta - y_1y_2\alpha_2^{next}\alpha_2 \\ +\alpha_2 y_2 z_2 + y_1 z_1(\beta - y_1 y_2 \alpha_2^{next}) + \alpha_2^{next}) + R(\alpha \setminus \{\alpha_1, \alpha_2\}). \tag{1.46}$$

Setting the derivative of $Q(\alpha_2)$ to zero, we obtain

$$\frac{dQ(\alpha_2)}{d\alpha_2} = 0, \tag{1.47}$$

which gives

$$\alpha_2(K_{11} + K_{22} - 2K_{12}) - y_1y_2\beta(K_{11} - K_{22}) - y_2(z_1 - z_2) = 1 - y_1y_2. \tag{1.48}$$

In order for $Q(\alpha_2)$ to have a maximum at α_2^{next}, we require the second derivative of $Q(\alpha_2)$ to be negative at α_2^{next}, i.e.,

$$\frac{d^2Q(\alpha_2)}{d\alpha_2^2} < 0, \tag{1.49}$$

which implies

$$K_{11} + K_{22} - 2K_{12} < 0. \tag{1.50}$$

The update proceeds by assuming that α_2^{next} is a maximum, i.e. (1.50) holds. Some algebraic simplification leads us to the relation between α^2 and α_2^{next}

$$\alpha_2^{next} = \alpha_2 + \frac{1}{(K_{11} + K_{22} - 2K_{12})}[y_2(f(x^2) - f(x^1)) + (y_1 y_2 - 1)], \tag{1.51}$$

where

$$f(x) = \sum_{j=1}^{M} \alpha_j y_j K(x, x^j) + b. \tag{1.52}$$

Once the target value for α_2^{next} is computed using (1.51), we proceed to Step 3.

Step 3

The target value of α_2^{next} computed using (1.51) is now checked for the bound constraints, i.e. we clip α_2^{next} to α_{min} or α_{max} if these bounds are crossed. Finally, we obtain the updated value for α_1^{next} as

$$\alpha_1^{next} = \alpha_1 + y_1 y_2(\alpha_2 - \alpha_2^{new}). \tag{1.53}$$

Step 4

Go to Step 1 to check if there are any more violators.

The SMO algorithm is widely used in many SVM solvers, including LIBSVM (Chang and Lin [12]). LIBSVM in fact uses several additional heuristics to speed up convergence. However, more efficient approaches are possible, as we show in the sequel.

1.6 Other Approaches to Solving the SVM QPP

Small changes to the SVM formulation can in fact make it easier to solve. In Joshi et al. [13], the authors suggest the relaxed SVM and the relaxed LSSVM formulations.

1.6.1 The Relaxed SVM

Consider the following optimization problem in its primal form.

$$\begin{array}{ll} \underset{(\xi,w,b)}{\text{Min}} & \frac{1}{2}w^Tw + \frac{h}{2}b^2 + Ce^T\xi \\ \text{subject to} & \end{array}$$

$$\begin{aligned} y_k[w^T\phi(x^{(k)}) + b] \geq 1 - \xi_k, \\ \xi_k \geq 0, \quad (k = 1, 2, \ldots, m). \end{aligned} \tag{1.54}$$

The addition of $\frac{h}{2}b^2$, where h is positive constant, in the objective function distinguishes it from the classical SVM formulation. Note that the $h = 1$ case is similar to PSVM formulation where b^2 is also added in the objective function.

The Lagrangian for the problem (1.54) is given by

$$\begin{aligned} L = Ce^T\xi + \sum_{k=1}^{m}\alpha_k[1 - \xi_k - y_k(w^T\phi(x^{(k)}) + b)] + \frac{1}{2}(w^Tw + hb^2) \\ - \sum_{k=1}^{m}\beta_k\xi_k. \end{aligned} \tag{1.55}$$

The K.K.T optimality conditions are given by

$$\nabla_w L = 0 \Rightarrow w = \sum_{k=1}^{m}\alpha_k y_k\phi(x^{(k)}), \tag{1.56}$$

$$\frac{\partial L}{\partial b} = 0 \Rightarrow b = \frac{1}{h}\sum_{k=1}^{m}\alpha_k y_k, \tag{1.57}$$

$$\frac{\partial L}{\partial q_k} = 0 \Rightarrow C - \alpha_k - \beta_k = 0 \Rightarrow \alpha_k + \beta_k = C. \tag{1.58}$$

From (1.56) and (1.57), we observe that

$$w^T\phi(x) + b = \sum_{k=1}^{m}\alpha_k y_k\left[K(x^{(k)}, x) + \frac{1}{h}\right], \tag{1.59}$$

where the kernel K is defined in the usual manner.

With a little algebra, we obtain the dual problem as

$$\begin{array}{ll} \underset{\alpha}{\text{Min}} & \frac{1}{2}\sum_{i=1}^{m}\sum_{j=1}^{m}y_iy_j\alpha_i\alpha_j\left(K_{ij} + \frac{1}{h}\right) - \sum_{i=1}^{m}\alpha_i \\ \text{subject to} & \end{array}$$

$$0 \leq \alpha_i \leq C, \quad (i = 1, 2, \ldots, m). \tag{1.60}$$

We refer to (1.60) as the Relaxed SVM (RSVM) formulation.

1.6.2 The Relaxed LSSVM

Consider the optimization problem

$$\underset{(\xi, w)}{\text{Min}} \quad \frac{1}{2} w^T w + \frac{h}{2} b^2 + C \xi^T \xi$$
subject to

$$y_i [w^T \phi(x^{(i)}) + b] + \xi_i = 1, \quad (i = 1, 2, \ldots m). \tag{1.61}$$

The Lagrangian for the problem (1.61) is given by

$$L = \frac{1}{2}(w^T w + h b^2) + C \xi^T \xi + \sum_{k=1}^{m} \alpha_k [1 - \xi_k - y_k [w^T \phi(x^{(k)}) + b]]. \tag{1.62}$$

The K.K.T optimality conditions are given by

$$\nabla_w L = 0 \Rightarrow w = \sum_{k=1}^{m} \alpha_k y_k \phi(x^{(k)}), \tag{1.63}$$

$$\frac{\partial L}{\partial b} = 0 \Rightarrow b = \frac{1}{h} \sum_{k=1}^{m} \alpha_k y_k, \tag{1.64}$$

$$\frac{\partial L}{\partial \xi_k} = 0 \Rightarrow \xi_k = (\frac{\alpha_k}{C}) \, . \tag{1.65}$$

From (1.63) and (1.64), we observe that

$$w^T \phi(x) + b = \sum_{k=1}^{m} \alpha_k y_k \left[K(x^{(k)}, x) + \frac{1}{h} + \frac{1}{C} \right], \tag{1.66}$$

where the kernel K is defined in the usual manner.

On simplifying, we obtain the dual problem as

$$\underset{\alpha}{\text{Min}} \quad \frac{1}{2} \sum_{i=1}^{m} \sum_{j=1}^{m} y_i y_j \alpha_i \alpha_j \left(P_{ij} + \frac{1}{h} \right) - \sum_{i=1}^{m} \alpha_i, \tag{1.67}$$

where P_{ij} is given by

$$P_{ij} = \begin{cases} K_{ij}, & i \neq j, \\ K_{ii} + \frac{1}{C}, & i = j. \end{cases} \tag{1.68}$$

1.6.3 *Solving the Relaxed SVM and the Relaxed LSSVM*

The duals of the RSVM and RLSSVM may be written in a common manner as

$$\underset{\lambda}{\text{Min}} \ \frac{1}{2}\sum_{i=1}^{m}\sum_{j=1}^{m} y_i y_j \lambda_i \lambda_j \left(Q_{ij}\right) - \sum_{i=1}^{m} \lambda_i. \tag{1.69}$$

Here $Q_{ij} = K_{ij} + \alpha_p$ in case of RSVM and $Q_{ij} = P_{ij} + \alpha_p$ in case of RLSSVM, where P_{ij} is given by (1.68). Box constraints are present in the case of the RSVM only. Since the linear constraint of the form (1.17) is absent from these formulations, we can develop an algorithm to update multipliers one by one. This is what the 1SMO algorithm achieves, as proposed by Joshi et al. [13].

Consider the problem

$$\underset{\alpha}{\text{Min}} \ \frac{1}{2}\sum_{i=1}^{m}\sum_{j=1}^{m} y_i y_j \alpha_i \alpha_j \left(Q_{ij}\right) - \sum_{i=1}^{m} \alpha_i. \tag{1.70}$$

Note that the boundary constraints

$$0 \leq \alpha_i \leq C, \quad (i = 1, 2, \ldots, m), \tag{1.71}$$

may be kept in mind when updating the multipliers. Without loss of generality, let α_1 be the multiplier under consideration. The objective function in (1.70) may be written as

$$\begin{aligned} Q(\alpha_1) = -\alpha_1 - \sum_{j=2}^{m} \alpha_j + \alpha_1 y_1 \sum_{j=2}^{m} \alpha_j y_j (Q_{1j} + \frac{1}{2}\alpha_1^2 (Q_{11}) \\ + \frac{1}{2}\sum_{i=2}^{m}\sum_{j=2}^{m} y_i y_j \alpha_i \alpha_j (Q_{ij}). \end{aligned} \tag{1.72}$$

We assume Q to be symmetric. Note that $y_1^2 = 1$ for (1.72). For the new value of α_1 to be a minimum of $Q(\alpha_1)$, we have $\frac{\partial Q}{\partial \alpha_1} = 0$, which gives

$$1 - \alpha_1^{new}(Q_{11}) - y_1 \sum_{j=2}^{m} \alpha_j y_j (K_{1j}) = 0. \tag{1.73}$$

We also require

$$\frac{\partial^2 Q}{\partial \alpha_1^2} > 0, i.e., Q_{11} > 0. \tag{1.74}$$

From (1.73) we obtain

$$\alpha_1^{new}(Q_{11}) = 1 - y_1 \sum_{j=2}^{m} \alpha_j y_j (Q_{1j}). \tag{1.75}$$

Denote $f(x) = w^T \phi(x) + b$. We therefore have

$$\alpha_1^{new}(Q_{11}) = \alpha_1^{old}(Q_{11}) + 1 - y_1 f^{old}(x^{(1)}), \tag{1.76}$$

which simplifies to

$$\alpha_k^{new} = \alpha_k^{old} + \frac{1 - y_k f^{old}(x^{(k)})}{(Q_{kk})}, \tag{1.77}$$

where the update rule has been changed to the case for any generic multiplier α_k. The boundary constraints (1.71) are enforced when updating the multipliers. The update rule (1.77) may be written as

$$\alpha_k^{new} = \alpha_k^{old} + \frac{1 - y_k f^{old}(x^{(k)})}{(K_{kk} + \alpha_p)}, \tag{1.78}$$

in the case of the relaxed SVM, and

$$\alpha_k^{new} = \alpha_k^{old} + \frac{1 - y_k f^{old}(x^{(k)})}{(P_{kk} + \alpha_p)}, \tag{1.79}$$

in the case of the relaxed LSSVM, where P_{ij} is given by

$$P_{ij} = \begin{cases} K_{ij}, & i \neq j, \\ K_{ii} + \frac{1}{C}, & i = j. \end{cases} \tag{1.80}$$

Note that the 1SMO update rules do not necessitate updating pairs of multipliers as in the case of the SMO [11]. This leads to faster updates, as shown in Joshi et al. [13].

Remark 1.6.1 Keerthi and Shevade [14], and Shevade et al. [15] discussed certain improvements in Platt's SMO algorithm for SVM classifier design and regression.

Keerthi and Shevade [14] also developed SMO algorithm for least squares SVM formulations. Some other efficient algorithms for solving SVM type formulations are Successive Over relaxation (SOR) due to Mangasarian and Musicant [16] and Iterative Single Data Algorithm due to Vogt et al. [17]. The SOR algorithm of Mangasarian and Musicant [16] has also been used in the study of non-parallel type classification, e.g., Tian et al. [18] and Tian et al. [19].

1.7 Conclusions

The main focus of this chapter has been to present an overview of Support Vector Machines and some of its variants, e.g., Least Squares Support Vector Machine and Proximal Support Vector Machines. Support Vector Regression (SVR) has been discussed in a totally different perspective. This development of SVR is based on a very important result of Bi and Bennett [4] which shows that the regression problem is equivalent to an appropriately constructed classification problem in R^{n+1}. Some popular and efficient algorithms for solving SVM problems have also been included in this chapter.

References

1. Vapnik, V. (1998). *Statistical learning theory*. New York: Wiley.
2. Suykens, J. A., & Vandewalle, J. (1999). Least squares support vector machine classifiers. *Neural Processing Letters*, *9*(3), 293–300.
3. Fung, G., & Mangasarian, O. L. (2001). Proximal support vector machine classifiers. In F. Provost & R. Srikant (Eds.), *Proceedings of Seventh International Conference on Knowledge Discovery and Data Mining* (pp. 77–86).
4. Bi, J., & Bennett, K. P. (2003). A geometric approach to support vector regression. *Neurocomputing*, *55*, 79–108.
5. Burges, C. (1998). A tutorial on support vector machines for pattern recognition. *Data Mining and Knowledge Discovery*, *2*(2), 121–167.
6. Sastry, P. (2003). An introduction to support vector machines. *Computing and information sciences: Recent trends*. New Delhi: Narosa Publishing House.
7. Deng, N., Tian, Y., & Zhang, C. (2012). *Support vector machines: Optimization based theory, algorithms and extensions*. New York: Chapman & Hall, CRC Press.
8. Chandra, S., Jayadeva, & Mehra, A. (2009). *Numerical optimization with applications*. New Delhi: Narosa Publishing House.
9. Mangasarian, O. L. (1994). *Nonlinear programming*. Philadelphia: SIAM.
10. Mangasarian, O. L., & Wild, E. W. (2006). Multisurface proximal support vector machine classification via generalized eigenvalues. *IEEE Transactions on Pattern Analysis and Machine Intelligence*, *28*(1), 69–74.
11. Platt, J. C. (1999). Fast training of support vector machines using sequential minimal optimization. In B. Schölkopf, C. Burges, & A. Smola (Eds.), *Advances in kernel methods: Support vector learning* (pp. 185–208). Cambridge: MIT Press.
12. Chang, C. C., & Lin, C. C. (2011). LIBSVM: A library for support vector machines. *ACM Transactions on Intelligent Systems and Technology (TIST)*, *2*(27). doi:10.1145/1961189.1961199.

13. Joshi, S., Jayadeva, Ramakrishanan, G., & Chandra, S. (2012). Using sequential unconstrained minimization techniques to simplify SVM solvers. *Neurocomputing*, *77*, 253–260.
14. Keerthi, S. S., & Shevade, S. K. (2003). SMO algorithm for least squares SVM formulations. *Neural Computations*, *15*(2), 487–507.
15. Keerthi, S. S., Shevade, S. K., Bhattacharya, C., & Murthy, K. R. K. (2001). Improvements to Platt's SMO algorithm for SVM classifier design. *Neural Computations*, *13*(3), 637–649.
16. Mangasarian, O. L., & Musicant, D. R. (1988). Successive overrelaxation for support vector machines. *IEEE Transactions on Neural Networks*, *10*(5), 1032–1037.
17. Vogt, M., Kecman, V., & Huang, T. M. (2005). Iterative single data algorithm for training kernel machines from huge data sets: Theory and performance. *Support vector machines: Theory and Applications*, *177*, 255–274.
18. Tian, Y. J., Ju, X. C., Qi, Z. Q., & Shi, Y. (2013). Improved twin support vector machine. *Science China Mathematics*, *57*(2), 417–432.
19. Tian, Y. J., Qi, Z. Q., Ju, X. C., Shi, Y., & Liu, X. H. (2013). Nonparallel support vector machines for pattern classification. *IEEE Transactions on Cybernertics*, *44*(7), 1067–1079.

Chapter 2
Generalized Eigenvalue Proximal Support Vector Machines

2.1 Introduction

In Chap. 1, we have presented a brief account of Proximal Support Vector Machines (PSVM) for the binary data classification problem. The PSVM formulation determines two parallel planes such that each plane is closest to one of the two data sets to be classified and the planes are as far apart as possible. The final separating hyperplane is then selected as the one which is midway between these two parallel hyperplanes. In a later development, Mangasarian and Wild [1] presented a new formulation where the two planes are no longer required to be parallel. Here, each plane is required to be as close as possible to one of the data sets and as far as possible from the other data set. They termed this new formulation as Generalized Eigenvalue Proximal Support Vector Machine (GEPSVM) because the required (non-parallel) planes are determined by solving a pair of generalized eigenvalue problems. We are discussing GEPSVM type models in this chapter because the basic GEPSVM model of Mangasarian and Wild [1] has been the major motivation for the introduction of Twin Support Vector Machines.

This chapter consists of three main sections, namely, GEPSVM for Classification: Mangasarian and Wild's Model, Variants of GEPSVM for Classification, and GEPSVR: Generalized Eigenvalue Proximal Support Vector Regression. Our presentation here is based on Mangasarian and Wild [1], Guarracino et al. [2], Shao et al. [3] and Khemchandani et al. [4].

2.2 GEPSVM for Classification

Let the training set for the binary data classification be given by

$$T_C = \{(x^{(i)}, y_i), i = 1, 2, ..., m\}, \tag{2.1}$$

Jayadeva et al., *Twin Support Vector Machines*, Studies in Computational Intelligence 659, DOI 10.1007/978-3-319-46186-1_2

where $x^{(i)} \in \mathbb{R}^n$ and $y_i \in \{-1, 1\}$. Let there be m_1 patterns having class label $+1$ and m_2 patterns having class label -1. We construct matrix A (respectively B) of order $(m_1 \times n)$ (respectively $(m_2 \times n)$) by taking the $i\,th$ row of A (respectively B) as the $i\,th$ pattern of class label $+1$ (respectively class label -1). Thus, $m_1 + m_2 = m$.

2.2.1 Linear GEPSVM Classifier

The linear GEPSVM classifier aims to determine two non parallel planes

$$x^T w_1 + b_1 = 0 \quad \text{and} \quad x^T w_2 + b_2 = 0, \tag{2.2}$$

such that the first plane is closest to the points of class $+1$ and farthest from the points of class -1, while the second plane is closest to the points in class -1 and farthest for the points in class $+1$. Here $w_1, w_2 \in \mathbb{R}^n$, and $b_1, b_2 \in \mathbb{R}$.

In order to determine the first plane $x^T w_1 + b_1 = 0$, we need to solve the following optimization problem

$$\underset{(w,b)\neq 0}{Min} \quad \frac{\|Aw + eb\|^2/\|(w,b)^T\|^2}{\|Bw + eb\|^2/\|(w,b)^T\|^2}, \tag{2.3}$$

where e is a vector of ‘ones’ of appropriate dimension and $\|.\|$ denotes the L_2-norm.

In (2.3), the numerator of the objective function is the sum of the squares of two-norm distances between each of the points of class $+1$ to the plane, and the denominator is the similar quantity between each of the points of class -1 to the plane. The way this objective function is constructed, it meets the stated goal of determining a plane which is closest to the points of class $+1$ and farthest to the points of class -1. Here, it may be remarked that ideally we would like to minimize the numerator and maximize the denominator, but as the same (w, b) may not do the simultaneous optimization of both the numerator and denominator, we take the ratio and minimize the same. This seems to be a very natural motivation for introducing the optimization problem (2.3).

The problem (2.3) can be re-written as

$$\underset{(w,b)\neq 0}{\text{Min}} \quad \frac{\|Aw + eb\|^2}{\|Bw + eb\|^2}, \tag{2.4}$$

where it is assumed that $(w, b) \neq 0$ implies that $(Bw + eb) \neq 0$. This makes the problem (2.4) well defined. Now, let

$$G = [A \quad e]^T [A \quad e],$$
$$H = [B \quad e]^T [B \quad e],$$

and $z = (w, b)^T$. Then (2.4) becomes

$$\underset{z \neq 0}{\text{Min}} \ \frac{z^T G z}{z^T H z}. \tag{2.5}$$

The objective function in (2.5) is the famous Rayleigh quotient of the generalized eigenvalue problem $Gz = \lambda Hz$, $z \neq 0$. When H is positive definite, the Rayleigh quotient is bounded and its range is $[\lambda_{min}, \lambda_{max}]$ where λ_{min} and λ_{max} respectively denote the smallest and the largest eigenvalues. Here, G and H are symmetric matrices of order $(n+1) \times (n+1)$, and H is positive definite under the assumption that columns of $[B \quad e]$ are linearly independent.

Now following similar arguments, we need to solve the following optimization problem to determine the second plane $x^T w_2 + b_2 = 0$

$$\underset{(w,b) \neq 0}{Min} \ \frac{\|Bw + eb\|^2}{\|Aw + eb\|^2}, \tag{2.6}$$

where we need to assume that $(w, b) \neq 0$ implies that $(Aw + eb) \neq 0$. We can again write (2.6) as

$$\underset{z \neq 0}{Min} \ \frac{z^T H z}{z^T G z}, \tag{2.7}$$

where G can be assumed to be positive definite provided columns of $[A \quad e]$ are linearly independent.

But in general, there is always a possibility that the null spaces of G and H have a nontrivial intersection. In that case the generalized eigenvalue problems (2.5) and (2.7) become singular and hence require a regularization technique to render their solutions. In this context, we may refer to Parlett [5] where the symmetric eigenvalue problems are discussed in greater detail.

Mangasarian and Wild [1] proposed the introduction of a Tikhonov type regularization (Tikhonov and Arsen [6]) in problems (2.4) and (2.6) to get

$$\underset{(w,b) \neq 0}{\text{Min}} \ \frac{\|Aw + eb\|^2 + \delta \|(w, b)^T\|^2}{\|Bw + eb\|^2}, \tag{2.8}$$

and

$$\underset{(w,b) \neq 0}{\text{Min}} \ \frac{\|Bw + eb\|^2 + \delta \|(w, b)^T\|^2}{\|Aw + eb\|^2}, \tag{2.9}$$

respectively. Here $\delta > 0$. Further a solution of (2.8) gives the first plane $x^T w_1 + b_1 = 0$ while that of (2.9) gives the second plane $x^T w_2 + b_2 = 0$.

Let us now introduce the following notations

$$\begin{aligned} P &= [A \quad e]^T[A \quad e] + \delta I, \\ Q &= [B \quad e]^T[B \quad e], \\ R &= [B \quad e]^T[B \quad e] + \delta I, \\ S &= [A \quad e]^T[A \quad e], \\ z^T &= (w, b), \end{aligned} \tag{2.10}$$

and rewrite problems (2.8) and (2.9) as

$$\underset{z \neq 0}{Min} \ \frac{z^T P z}{z^T Q z}, \tag{2.11}$$

and

$$\underset{z \neq 0}{Min} \ \frac{z^T R z}{z^T S z}, \tag{2.12}$$

respectively.

We now have the following theorem of Parlett [5].

Theorem 2.2.1 (Parlett [5])

Let E and F be arbitrary $(n \times n)$ real symmetric matrices and F be positive definite. Let for $u \neq 0$, $\gamma(u) = ((u^T Eu)/(u^T Fu))$ be the Rayleigh quotient and $Eu = \theta Fu$ be the corresponding generalized eigenvalue problem. Then

1. *$\gamma(u)$ ranges over $[\theta_{min}, \theta_{max}]$ where θ_{min} and θmax are the smallest and the largest eigenvalues of the generalized eigenvalue problem $Eu = \theta Fu$.*
2. *If $\bar{u}$ is a stationary point of $\gamma(u)$, then $\bar{u}$ is an eigenvector of the corresponding generalized eigenvalue problem $Eu = \theta Fu$ and conversely.*

In view of Theorem 2.2.1, solving the optimization problem (2.11) amounts to finding the eigenvector corresponding to the smallest eigenvalue of the generalized eigenvalue problem

$$Pz = \lambda Qz, \quad z \neq 0. \tag{2.13}$$

In a similar manner, to solve the optimization problem (2.12) we need to get the eigenvector corresponding to the smallest eigenvalue of the generalized eigenvalue problem

$$Rz = \mu Sz, \quad z \neq 0. \tag{2.14}$$

We can summarize the above discussion in the form of the following theorem.

Theorem 2.2.2 (Mangasarian and Wild [1])

Let A and B be $(m_1 \times n)$ and $(m_2 \times n)$ matrices corresponding to data sets of class $+1$ and class -1 respectively. Let P, Q, R and S be defined as at (2.10). Let columns of $(A \quad e)$ and $(B \quad e)$ be linearly independent. Let z_1 be the eigenvector corresponding to smallest eigenvalue of the generalized eigenvalue problem (2.13). Also, let z_2 be the eigenvector corresponding to the smallest eigenvector of the generalized eigenvalue problem (2.14). Then, $z_1 = col(w_1, b_1)$ and $z_2 = col(w_2, b_2)$.

Remark 2.2.1 z_1 and z_2 determine the planes $x^T w_1 + b_1 = 0$ and $x^T w_2 + b_2 = 0$. Further, the generalized eigenvalue problems (2.13) and (2.14) can be solved very easily by employing two MATLAB commands: $eig(P, Q)$ and $eig(R, S)$ respectively.

Remark 2.2.2 A new point $x \in \mathbb{R}^n$ is assigned to the class i, $(i = 1, 2)$ depending on which of the two planes this point is closest to, i.e. class $= \underset{i=1,2}{arg\ min}\{|x^T w_i + b_i|/\|w_i\|^2\}$.

Remark 2.2.3 The requirement that columns of $(A \quad e)$ and $(B \quad e)$ are linearly independent, is only a sufficient condition for the determination of z_1 and z_2. It is not a necessary condition as may be verified for the XOR example. Also this linear independence condition may not be too restrictive if m_1 and m_2 are much larger than n.

We now discuss a couple of examples for illustration purposes.

Example 2.2.1 (XOR Problem)

The training set for the XOR problem is given by

$$T_C = \{((0, 0), +1), ((1, 1), +1), ((1, 0), -1), ((0, 1), -1).\}$$

Therefore

$$A = \begin{bmatrix} 0 & 0 \\ 1 & 1 \end{bmatrix} \quad \text{and} \quad B = \begin{bmatrix} 1 & 0 \\ 0 & 1 \end{bmatrix}.$$

Hence

$$P = S = \begin{bmatrix} 1 & 1 & 1 \\ 1 & 1 & 1 \\ 1 & 1 & 2 \end{bmatrix},$$

and

$$Q = R = \begin{bmatrix} 1 & 0 & 1 \\ 0 & 1 & 1 \\ 1 & 1 & 2 \end{bmatrix}.$$

Then the generalized eigenvalue problem (2.13) has the solution $\lambda_{min} = 0$ and $z_1 = (-1 \quad 1 \quad 0)$. Therefore, the first plane is given by $-x_1 + x_2 + 0 = 0$, i.e., $x_1 - x_2 = 0$. In the same way, the generalized eigenvalue problem (2.14) has the solution $\mu_{min} = 0$ and $z_2 = (-1 \quad -1 \quad 1)$, which gives the second plane $-x_1 - x_2 + 1 = 0$, i.e., $x_1 + x_2 = 1$.

Here we observe that neither the columns of $(A \quad e)$ nor that of $(B \quad e)$ are linearly independent but the problems (2.13) and (2.14) have solutions z_1 and z_2 respectively.

The XOR example also illustrates that proximal separability does not imply linear separability. In fact these are two different concepts and therefore the converse is also not true. It is also possible that two sets be both proximally and linearly separable, e.g. the AND problem.

Example 2.2.2 (*Cross Planes Example : Mangasarian and Wild* [1])

Mangasarian and Wild [1] constructed the famous Cross Planes example, where the data consists of points that are close to one of the two intersecting cross planes in $\mathbb{R}^2$. The below given Fig. 2.1 depicts the Cross Planes data sets in $\mathbb{R}^2$. Here it can be observed that for this data set, the training set correctness of GEPSVM is 100 %, while it is only 80 % for the linear PSVM classifier.

Geometrically, the Cross Planes data set is a perturbed generalization of the XOR example. Similar to XOR, it serves as a test example for the efficacy of typical linear classifiers, be it linear SVM, linear PSVM, linear GEPSVM or others to be studied in the sequel. An obvious reason for the poor performance of linear PSVM on Cross Planes data set is that in PSVM we have insisted on the requirement that the proximal planes should be parallel, which is not the case with linear GEPSVM.

We next discuss the kernel version of GEPSVM to get the nonlinear GEPSVM classifier.

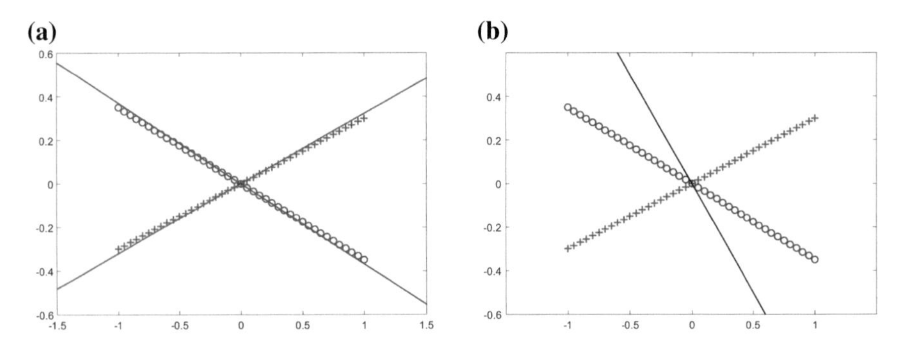

Fig. 2.1 Cross planes dataset: **a** GEPSVM classifier **b** SVM classifier

2.2.2 Nonlinear GEPSVM Classifier

Let $C^T = [A \quad B]^T$ and K be an appropriate chosen kernel. We wish to construct the following two kernel generated surfaces

$$K(x^T, C^T)u_1 + b_1 = 0, \quad \text{and} \quad K(x^T, C^T)u_2 + b_2 = 0. \tag{2.15}$$

Here we note that (2.15) are nonlinear surfaces rather than planes, but serve the same purpose as (2.2). Thus, the first (respectively second) surface is closest to data points in class $+1$ (respectively class -1) and farthest from the data points in class -1 (respectively class $+1$).

If we take the linear kernel $K(x^T, C^T) = x^T C$ and define $w_1 = C^T u_1$, $w_2 = C^T u_2$, then (2.15) reduces to (2.2). Therefore to generate surfaces (2.15) we can generalize our earlier arguments and get the following two optimization problems

$$\underset{(u,b)\neq 0}{Min} \quad \frac{\|K(A, C^T)u + eb\|^2 + \delta\|(u, b)^T\|^2}{\|K(B, C^T)u + eb\|^2}, \tag{2.16}$$

and

$$\underset{(u,b)\neq 0}{Min} \quad \frac{\|K(B, C^T)u + eb\|^2 + \delta\|(u, b)^T\|^2}{\|K(A, C^T)u + eb\|^2}. \tag{2.17}$$

If we now define

$$\begin{aligned}
P_1 &= [K(A, C^T) \quad e]^T[K(A, C^T) \quad e] + \delta I,\\
Q_1 &= [K(B, C^T) \quad e]^T[K(B, C^T) \quad e],\\
R_1 &= [K(B, C^T) \quad e]^T[K(B, C^T) \quad e] + \delta I,\\
S_1 &= [K(A, C^T) \quad e]^T[K(A, C^T) \quad e],\\
z^T &= (u, b),
\end{aligned} \tag{2.18}$$

then the optimization problem (2.16) and (2.17) reduces to generalized eigenvalue problems

$$\underset{z\neq 0}{Min} \quad \frac{z^T P_1 z}{z^T Q_1 z}, \tag{2.19}$$

and

$$\underset{z\neq 0}{Min} \quad \frac{z^T R_1 z}{z^T S_1 z}, \tag{2.20}$$

respectively.

Now we can state a theorem similar to Theorem (2.2.2) and make use of the same to generate the required surfaces (2.15). Specifically, let z_1 (respectively z_2) be the eigenvector corresponding to the smallest eigenvalue of the generalized eigenvalue problem (2.19) and (respectively 2.20), the $z_1 = (u_1, b_1)$ (respectively $z_2 = (u_2, b_2)$) gives the desired surface $K(x^T, C^T)u_1 + b_1 = 0$ (respectively $K(x^T, C^T)u_2 + b_2 = 0$.). For a new point $x \in \mathbb{R}^n$, let

$$\text{distance}(x, S_i) = \frac{|K(x^T, C^T)u_i + b_i|}{\sqrt{(u_i)^T K(C, C^T)u_i}},$$

where S_i is the non linear surface $K(x^T, C^T)u_i + b_i = 0$ $(i = 1, 2)$. Then the class i $(i = 1, 2)$ for this new point x is assigned as per the following rule

$$\text{class} = \underset{i=1,2}{arg\,Min} \frac{|K(x^T, C^T)u_i + b_i|}{\sqrt{(u_i)^T K(C, C^T)u_i}}. \tag{2.21}$$

Mangasarian and Wild [1] implemented their linear and nonlinear GEPSVM classifiers extensively on artificial as well as real world data sets. On Cross Planes data set in (300×7) dimension (i.e. $m = 300, n = 7$) the linear classifier gives 10-fold testing correctness of 98 % where as linear PSVM and linear SVM give the correctness of 55.3 % and 45.7 % respectively. Also on the Galaxy Bright data set, the GEPSVM linear classifier does significantly better than PSVM. Thus, Cross Planes as well as Galaxy Bright data sets indicate that allowing the proximal planes to be non-parallel allows the classifier to better represent the data set when needed. A similar experience is also reported with the nonlinear GEPSVM classifier on various other data sets.

2.3 Some Variants of GEPSVM for Classification

There are two main variants of original GEPSVM. These are due to Guarracino et al. [2] and Shao et al. [3]. Guarracino et al. [2] proposed a new regularization technique which results in solving a single generalized eigenvalue problem. This formulation is termed as Regularized GEPSVM and is denoted as ReGEPSVM. The formulation of Shao et al. [3] is based on the difference measure, rather than the ratio measure of GEPSVM, and therefore results in solving two eigenvalue problems, unlike the two generalized eigenvalue problems in GEPSVM. Shao et al. [3] model is termed as Improved GEPSVM and is denoted as IGEPSVM. Both of these variants are attractive as they seem to be superior to the classical GEPSVM in terms of classification accuracy as well as in computation time. This has been reported in [2] and [3] by implementing and experimenting these algorithms on several artificial and benchmark data sets.

2.3.1 ReGEPSVM Formulation

Let us refer to the problems (2.8) and (2.9) where Tikhonov type regularization term is introduced in the basic formulation of (2.4) and (2.6). Let us also recall the following theorem from Saad [7] in this regard.

Theorem 2.3.1 *Let τ_1, τ_2, δ_1 and δ_2 be scalars such that $(\tau_1\tau_2 - \delta_1\delta_2) \neq 0$. Let $G^* = (\tau_1 G - \delta_1 H)$ and $H^* = (\tau_2 H - \delta_2 G)$. Then the generalized eigenvalue problems $G^*x = \lambda H^*x$ and $Gx = \mu Hx$ are related as follows*

1. *if $\overline{\lambda}$ is an eigenvalue of $G^*x = \lambda H^*x$ and $\overline{\mu}$ is an eigenvalue of $Gx = \mu Hx$, then*

$$\overline{\mu} = \frac{\tau_2\overline{\lambda} + \delta_1}{\tau_1 + \delta_2\overline{\lambda}},$$

2. *both generalized eigenvalue problems have the same eigenvectors.*

We now consider the following optimization problem

$$\underset{(w,b)\neq 0}{Min} \ \frac{\|Aw + eb\|^2 + \hat{\delta_1}\|Bw + eb\|^2}{\|Bw + eb\|^2 + \hat{\delta_2}\|Aw + eb\|^2}, \tag{2.22}$$

which is similar to (2.8) (and also to (2.9)) but with a different regularization term. Now if we take $\tau_1 = \tau_2 = 1$, $\hat{\delta_1} = -\delta_1 > 0$, $\hat{\delta_2} = -\delta_2 > 0$ with the condition that $\tau_1\tau_2 - \delta_1\delta_2 \neq 0$, then Theorem (2.3.1) becomes applicable for the generalized eigenvalue problem corresponding to the optimization problem (2.22). Let

$$U = [A \quad e]^T[A \quad e] + \hat{\delta_1}[B \quad e]^T[B \quad e],$$
$$V = [B \quad e]^T[B \quad e] + \hat{\delta_2}[A \quad e]^T[A \quad e].$$

Then, the generalized eigenvalue problem corresponding to (2.22) is

$$Uz = \lambda Vz, \qquad z \neq 0. \tag{2.23}$$

Further, the smallest eigenvalue of the original problem (2.5) becomes the largest eigenvalue of (2.23), and the largest eigenvalue of (2.5) becomes the smallest eigenvalue of (2.23). This is because of Theorem (2.3.1) which asserts that the spectrum of the transformed eigenvalue problem $G^*x = \lambda H^*x$ gets shifted and inverted. Also, the eigenvalues of (2.5) and (2.7) are reciprocal to each other with the same eigenvectors. Therefore, to determine the smallest eigenvalues of (2.5) and (2.7) respectively, we need to determine the largest and the smallest eigenvalues of (2.23). Let $z_1 = col(w_1, b_1)$ and $z_2 = col(w_2, b_2)$ be the corresponding eigenvectors. Then, the respective planes are $x^T w_1 + b_1 = 0$ and $x^T w_2 + b_2 = 0$.

For the nonlinear case, we need to solve the analogous optimization problem

$$\underset{(u,b)\neq 0}{Min} \quad \frac{\|K(A, C^T)u + eb\|^2 + \hat{\delta}_1\|K(B, C^T)u + eb\|^2}{\|K(B, C^T)u + eb\|^2 + \hat{\delta}_2\|K(A, C^T)u + eb\|^2}, \tag{2.24}$$

via the related generalized eigenvalue problem $G^*x = \lambda H^*x$. Its solution will give two proximal surfaces

$$K(x^T, C^T)u_1 + b_1 = 0 \quad \text{and} \quad K(x^T, C^T)u_2 + b_2 = 0,$$

and the classification for a new point $x \in \mathbb{R}^n$ is done similar to nonlinear GEPSVM classifier.

2.3.2 Improved GEPSVM Formulation

It has been remarked in Sect. 2.1 that the determination of the first plane $x^T w_1 + b_1 = 0$ (respectively the second plane $x^T w_2 + b_2 = 0$) requires the minimization (respectively maximization) of $(\|Aw + eb\|^2/\|(w, b)^T\|^2)$ and the maximization (respectively minimization) of $(\|Bw + eb\|^2/\|(w, b)^T\|^2)$. Since the same (w, b) may not perform this simultaneous optimization, Mangasarian and Wild [1] proposed a ratio measure to construct the optimization problem (2.4) (respectively (2.6)) and thereby obtained the generalized eigenvalue problem (2.13) (respectively (2.14)) for determining the first (respectively second) plane.

The main difference between the works of Mangasarian and Wild [1], and that of Shao et al. [3] is that the former considers a ratio measure whereas the latter considers a difference measure to formulate their respective optimization problems. Conceptually there is a bi-objective optimization problem which requires the minimization of $(\|Aw + eb\|^2/\|(w, b)^T\|^2)$ and maximization of $(\|Bw + eb\|^2/\|(w, b)^T\|^2)$, i.e. minimization of -$(\|Bw + eb\|^2/\|(w, b)^T\|^2)$. Thus the relevant optimization problem is the following bi-objective optimization problem

$$\underset{(w,b)\neq 0}{Min} \left(\frac{\|Aw + eb\|^2}{\|(w, b)^T\|^2}, -\frac{\|Bw + eb\|^2}{\|(w, b)^T\|^2} \right). \tag{2.25}$$

Let $\gamma > 0$ be the weighting factor which determines the trade-off between the two objectives in (2.25). Then, the bi-objective optimization problem (2.25) is equivalent to the following scalar optimization problem

$$\underset{(w,b)\neq 0}{Min} \left(\frac{\|Aw + eb\|^2}{\|(w, b)^T\|^2} - \gamma \frac{\|Bw + eb\|^2}{\|(w, b)^T\|^2} \right). \tag{2.26}$$

Let $z = (w, b)^T$, $G = [A \quad e]^T[A \quad e]$ and $H = [B \quad e]^T[B \quad e]$. Then, the optimization problem (2.26) becomes

$$\underset{z \neq 0}{Min} \; \frac{z^T (G - \gamma H) z}{z^T I z}. \tag{2.27}$$

Now similar to GEPSVM, we can also introduce a Tikhonov type regularization term (2.27) to get

$$\underset{z \neq 0}{Min} \; \frac{z^T (G - \gamma H) z + \delta \|z\|^2}{z^T I z},$$

i.e.

$$\underset{z \neq 0}{Min} \; \frac{z^T (G + \delta I - \gamma H) z}{z^T I z}. \tag{2.28}$$

The above problem (2.28) is exactly the minimization of the Rayleigh quotient whose global optimum solution can be obtained by solving the following eigenvalue problem

$$(G + \delta I - \gamma H) z = \lambda z, \qquad z \neq 0. \tag{2.29}$$

By determining the eigenvector corresponding to the smallest eigenvalue of (2.29) we get $z_1 = (w_1, b_1)^T$ and hence the desired first plane $x^T w_1 + b_1 = 0$.

In a similar manner, the second plane $x^T w_2 + b_2 = 0$ is obtained by determining the eigenvector $z_2 = (w_2, b_2)^T$ corresponding to the smallest eigenvalue for the eigenvalue problem

$$(H + \delta I - \gamma G) z = \mu z, \qquad z \neq 0. \tag{2.30}$$

In conclusion, the two desired planes $x^T w_i + b_i = 0, \; (i = 1, 2)$ can be obtained by solving two eigenvalue problems (2.29) and (2.30). Then as before, a new point $x \in \mathbb{R}^n$ is assigned to the class $i \; (i = 1, 2)$, depending on which of the two hyperplanes it is closer to, i.e.,

$$\text{class(x)} = \underset{i=1,2}{arg\,Min} \frac{\|x^T w_i + b_i\|}{\|w_i\|}.$$

The above results can easily be extended to the nonlinear case by considering the eigenvalue problems

$$(M + \delta I - \gamma N) z = \lambda z, \qquad z \neq 0, \tag{2.31}$$

and

$$(N + \delta I - \gamma M) z = \mu z, \qquad z \neq 0, \tag{2.32}$$

where

$$\begin{aligned} M &= [K(A, C^T) \quad e]^T [K(A, C^T) \quad e], \\ N &= [K(B, C^T) \quad e]^T [K(B, C^T) \quad e], \\ C^T &= [A \quad B]^T, \end{aligned}$$

and I is an identity matrix of appropriate dimension.

Let $z_1 = (u_1, b_1)^T$ (respectively $z_2 = (u_2, b_2)^T$) be the eigenvector corresponding to the smallest eigenvalue of (2.31) (respectively (2.32)) then the desired surfaces are $K(x^T, C^T)u_1 + b_1 = 0$ and $K(x^T, C^T)u_2 + b_2 = 0$ respectively. For a new point $x \in \mathbb{R}^n$ we again decide class label based on the following decision rule

$$\text{class(x)} = \underset{i=1,2}{arg\,Min} \frac{|K(x^T, C^T)u_i + b_i|}{\sqrt{(u_i)^T K(C, C^T)u_i}}. \tag{2.33}$$

Shao et al. [3] termed their model as Improved Generalized Eigenvalue Proximal Support Vector Machine (IGEPSVM). It is known that the linear IGEPSVM needs only to solve two eigenvalue problems with computational time complexity of $\mathcal{O}(n^2)$, where n is the dimensionality of data points. In contrast, the linear GEPSVM requires two generalized eigenvalue problems whose complexity is $\mathcal{O}(n^3)$. For the nonlinear case, the computational complexity is $\mathcal{O}(m^2)$ for Shao et al. model [3] and is $\mathcal{O}(m^3)$ for Mangasarian and Wild model. Here, m is the number of training data points. This probably explains that in numerical implementations, IGEPSVM takes much less time than GEPSVM. For details of these numerical experiments we shall refer to Shao et al. [3]. Recently, Saigal and Khemchandani [8] presented comparison of various nonparallel algorithms including variations of GEPSVM along with TWSVM for multicategory classification.

2.4 GEPSVR: Generalized Eigenvalue Proximal Support Vector Regression

The aim of this section is to present the regression problem in the setting of the generalized eigenvalue problem. Here, we present two models for the regression problem. The first is termed as the GEPSVR which is in the spirit of GEPSVM and requires the solution of two generalized eigenvalue problems. The second formulation is in the spirit of ReGEPSVM and is termed as the Regularized Generalized Eigenvalue Support Vector Regressor (ReGEPSVR). The formulation of ReGEPSVR requires the solution of a single regularized eigenvalue problem and it reduces the execution time to half as compared to GEPSVR.

Earlier Bi and Bennett [9] made a very significant theoretical contribution to the theory of support vector regression. They (Bi and Bennett [9]) showed that the problem of support vector regression can be regarded as a classification problem in the dual space and maximizing the margin corresponds to shrinking of effective ϵ-tube. From application point of view, this result is of utmost importance because this allows to look for other classification algorithms and study their regression analogues. We have already demonstrated this duality aspect in the context of SVR in Chap. 1, where it was derived via SVM approach. We shall continue to take this approach in this section as well as in some of the later chapters where regression problem is further studied.

2.4.1 GEPSVR Formulation

For GEPSVR, our objective is to find the two non-parallel ϵ-insensitive bounding regressors. The two non-parallel regressors around the data points are derived by solving two generalized eigenvalue problems. We first discuss the case of linear GEPSVR, the case of nonlinear GEPSVR may be developed analogously.

Let the training set for the regression problem be given by

$$T_R = \{(x^{(i)}, y_i), i = 1, 2, ..., m\},$$

for some $\epsilon > 0$ where $x^{(i)} \in \mathbb{R}^n$ and $y_i \in \mathbb{R}$.

Following Bi and Bennett [9], we consider the associated classification problem with the training set T_C in $\mathbb{R}^{n+1}$ given by

$$T_C = \{((x^{(i)}, y_i + \epsilon), +1), ((x^{(i)}, y_i - \epsilon), -1), i = 1, 2, ..., m\}, \text{ where } \quad \epsilon > 0.$$

The GEPSVR algorithm finds two functions $f_1(x) = x^T w_1 + b_1$ and $f_2(x) = x^T w_2 + b_2$ such that each one determines the ϵ-insensitive bounded regressor.

Let A be an $(m \times n)$ matrix whose $i\,th$ row is the vector $(x^{(i)})^T$. Let $(Y + \epsilon e) = (y_1 + \epsilon, ..., y_m + \epsilon)$ and $(Y - \epsilon e) = (y_1 - \epsilon, ..., y_m - \epsilon)$. Then $f_1(x) = x^T w_1 + b_1$ may be identified as a hyperplane in $\mathbb{R}^{n+1}$. The GEPSVR formulation treats the given regression problem as a classification problem in $\mathbb{R}^{n+1}$ with the training data set as T_C. Thus, $f_1(x)$ is determined so as to minimize the Euclidean distance of the hyperplane $f_1(x) = x^T w_1 + b_1$ in $\mathbb{R}^{n+1}$ from the set of points $(A, Y - \epsilon e)$ and maximize its Euclidean distance from the set of points $(A, Y + \epsilon e)$. Likewise $f_2(x)$ is determined so as to minimize the Euclidean distance of the hyperplane $f_2(x) = x^T w_2 + b_2$ in $\mathbb{R}^{n+1}$ from the set of points $(A, Y + \epsilon e)$ and maximize its distance from $(A, Y + \epsilon e)$.

This leads to the following optimization problem for determining $f_1(x)$, the optimization problem for determining $f_2(x)$ can be written on similar lines.

$$\underset{(w,b)\neq 0}{Min} \quad \frac{\|Aw + eb - (Y - e\epsilon)\|^2 / \|(w, b)^T\|^2}{\|Aw + eb - (Y + e\epsilon)\|^2 / \|(w, b)^T\|^2}. \tag{2.34}$$

Here, it is implicitly assumed that $(w, b) \neq 0$ implies that $(Aw + eb - (Y + e\epsilon)) \neq 0$. In that case (2.34) can be simplified as

$$\underset{(w,b)\neq 0}{Min} \quad \frac{\|Aw + eb - (Y - e\epsilon)\|^2}{\|Aw + eb - (Y + e\epsilon)\|^2}. \tag{2.35}$$

Now the optimization problem (2.35) can be regularized as

$$\underset{(w,b)\neq 0}{Min} \quad \frac{\|Aw + eb - (Y - e\epsilon)\|^2 + \delta\|(w \quad b \quad -1)^T\|^2}{\|Aw + eb - (Y + e\epsilon)\|^2}, \tag{2.36}$$

where $\delta > 0$ is the regularization coefficient. Let $\mu = (w \quad b \quad -1)^T$ and

$$R = [A \quad e \quad (Y - e\epsilon)]^T[A \quad e \quad (Y - e\epsilon)] + \delta I$$
$$S = [A \quad e \quad (Y + e\epsilon)]^T[A \quad e \quad (Y + e\epsilon)].$$

Then, the solution of regularized optimization problem (2.36) can be obtained by solving the following generalized eigenvalue problem

$$Ru = \eta Su, \qquad u \neq 0. \tag{2.37}$$

Let u_1 denote the eigenvector corresponding to the smallest eigenvalue η_{min} of (2.37). To obtain w_1 and b_1 from u_1, we normalize u_1 by the negative of the $(n+2)^{th}$ element of u_1 so as to force a (-1) at the $(n+2)^{th}$ position of u_1. Let this normalized representation of u_1 be u_1 with (-1) at the end, such that $u_1^{new} = [w_1 \quad b_1 \quad -1]^T$. Then, w_1 and b_1 determine an ϵ-insensitive bounding regressor as $f_1(x) = x^T w_1 + b_1$.

Similarly for determining the second bounding regressor $f_2(x)$ we consider the regularized optimization problem

$$\underset{(w,b)\neq 0}{Min} \quad \frac{\|Aw + eb - (Y + e\epsilon)\|^2 + \delta\|(w \quad b \quad -1)^T\|^2}{\|Aw + eb - (Y - e\epsilon)\|^2}, \tag{2.38}$$

where $(w, b) \neq 0$ implies $(Aw + eb - (Y + e\epsilon)) \neq 0$. Let $u = (w \quad b \quad -1)^T$ and

$$P = [A \quad e \quad (Y + e\epsilon)]^T[A \quad e \quad (Y + e\epsilon)] + \delta I,$$
$$Q = [A \quad e \quad (Y - e\epsilon)]^T[A \quad e \quad (Y - e\epsilon)].$$

Then, the optimization problem (2.38) is equivalent to the generalized eigenvalue problem

$$Pu = \nu Qu, \qquad u \neq 0. \tag{2.39}$$

Now as before, finding minimum eigenvalue ν_{min} of (2.37) and having determined the corresponding eigenvector u_2, we obtain w_2 and b_2 by the normalizing procedure explained earlier. This gives the other ϵ-insensitive regressor $f_2(x) = x^T w_2 + b_2$.

Having determined $f_1(x)$ and $f_2(x)$ from u_1 and u_2, the final regressor $f(x)$ is constructed by taking the average, i.e.,

$$f(x) = \frac{1}{2}(f_1(x) + f_2(x)) = \frac{1}{2}x^T(w_1 + w_2) + \frac{1}{2}(b_1 + b_2).$$

2.4.2 Regularized GEPSVR Formulation

Working on similar lines as that of ReGEPSVM, we intend to use regularization technique of Guarracino et al. [2] for finding a single regularized eigenvalue problem whose smallest and largest eigenvalue μ_{min} and μ_{max} would provide $(w_1, b_1)^T$ and $(w_2, b_2)^T$ from their corresponding eigenvectors. These solutions would then correspond to the two ϵ-insensitive bounding regressors $f_1(x) = x^T w_1 + b_1$ and $f_2(x) = x^T w_2 + b_2$ respectively. Let $\hat{\delta}_1 > 0$ and $\hat{\delta}_2 > 0$. We consider the following regularized optimization problem

$$\underset{(w,b)\neq 0}{Min} \quad \frac{\|Aw + eb - (Y - e\epsilon)\|^2 + \hat{\delta}_1\|Aw + eb - (Y + e\epsilon)\|^2}{\|Aw + eb - (Y + e\epsilon)\|^2 + \hat{\delta}_2\|Aw + eb - (Y - e\epsilon)\|^2}. \tag{2.40}$$

Let $t = [w \quad b \quad -1]^T$ and U and V be defined as

$$U = [A \quad e \quad (Y - e\epsilon)]^T[A \quad e \quad (Y - e\epsilon)] + \hat{\delta}_1[A \quad e \quad (Y + e\epsilon)]^T[A \quad e \quad (Y + e\epsilon)],$$
$$V = [A \quad e \quad (Y + e\epsilon)]^T[A \quad e \quad (Y + e\epsilon)] + \hat{\delta}_2[A \quad e \quad (Y - e\epsilon)]^T[A \quad e \quad (Y - e\epsilon)].$$

Now using the earlier discussed properties of Rayleigh quotient, the optimization problem (2.40) is equivalent to the following generalized eigenvalue problem

$$Ut = \nu Vt, \qquad t \neq 0. \tag{2.41}$$

This yields the eigenvector t_1 corresponding to largest eigenvalue ν_{max} of (2.41), and eigenvector t_2 corresponding to the smallest eigenvalue ν_{min} of (2.41). To obtain w_1 and b_1 from t_1 and w_2 and b_2 from t_2, we follow the usual normalization procedure of Sect. 2.4.1 and get $t_1^{new} = (w_1 \quad b_1 \quad -1)^T$ and $t_2^{new} = (w_2 \quad b_2 \quad -1)^T$. This yield the ϵ-insensitive bounding regressors $f_1(x) = x^T w_1 + b_1$ and $f_2(x) = x^T w_2 + b_2$ from t_1^{new} and t_2^{new}. For a new point $x \in \mathbb{R}^n$, the regressed value $f(x)$ is given by

$$f(x) = \frac{1}{2}(f_1(x) + f_2(x)) = \frac{1}{2}x^T(w_1 + w_2) + \frac{1}{2}(b_1 + b_2).$$

For extending our results to the nonlinear case, we consider the following kernel generated functions instead of linear functions

$$F_1(x) = K(x^T, A^T)w_1^{\phi} + b_1^{\phi} \quad \text{and} \quad F_2(x) = K(x^T, A^T)w_2^{\phi} + b_2^{\phi}, \tag{2.42}$$

where K is the chosen Kernel function and w_1^{ϕ}, w_2^{ϕ}, b_1^{ϕ} and b_2^{ϕ} are defined in the kernel spaces.
Let $t^{\phi} = [w^{\phi} \quad b^{\phi} \quad -1]^T$ and

$$\begin{aligned} E &= [K(A, A^T) \quad e \quad (Y - e\epsilon)]^T [K(A, A^T) \quad e \quad (Y - e\epsilon)] + \\ &\quad \hat{\delta}_1 [K(A, A^T) \quad e \quad (Y + e\epsilon)]^T [K(A, A^T) \quad e \quad (Y + e\epsilon)], \\ F &= [K(A, A^T) \quad e \quad (Y + e\epsilon)]^T [K(A, A^T) \quad e \quad (Y + e\epsilon)] + \\ &\quad \hat{\delta}_2 [K(A, A^T) \quad e \quad (Y - e\epsilon)]^T [K(A, A^T) \quad e \quad (Y - e\epsilon)]. \end{aligned}$$

We have the following generalized eigenvalue problem

$$Et^{\phi} = \beta F t^{\phi}, \qquad t^{\phi} \neq 0. \tag{2.43}$$

This yields the eigenvector t_1^{ϕ} corresponding to the largest eigenvalue β_{max} of (2.43), and t_2^{ϕ} corresponding to the smallest eigenvalue β_{min} of (2.43). We do the usual normalization of t_1^{ϕ} and t_2^{ϕ} to get $t_{new}^{\phi} = [w_1^{\phi} \quad b_1^{\phi} \quad -1]^T$, and $t_{new}^{\phi} = [w_2^{\phi} \quad b_2^{\phi} \quad -1]^T$. This gives the ϵ-insensitive bounding regressor $F_1(x) = K(x^T, A^T)w_1^{\phi} + b_1^{\phi}$ and $F_2(x) = K(x^T, A^T)w_2^{\phi} + b_2^{\phi}$. For a new input pattern $x \in \mathbb{R}^n$, the regressed value is given by

$$F(x) = \frac{1}{2}(w_1^{\phi} + w_2^{\phi})K(x^T, A^T) + \frac{1}{2}(b_1^{\phi} + b_2^{\phi}).$$

To test the performance of ReGEPSVR, Khemchandani et al. [4] implemented their model on several data sets including UCI, financial time series data and four two dimensional functions considered by Lázaro et al. [10]. Here, we summarize some of the conclusions reported in Khemchandani et al. [4].

2.4.3 Experimental Results

To test the performance of the proposed ReGEPSVR, Khemchandani et al. [4] compared it with SVR on several datasets. The performance of these regression algorithms on the above-mentioned datasets largely depends on the choice of initial parameters. Hence, optimal parameters for these algorithms for UCI datasets [11] are selected by using a cross-validation set [12] comprising of 10 percent of the dataset, picked up randomly. Further, RBF kernel is taken as the choice for kernel function in all the implementations, adding to another parameter, i.e. σ_{kernel} to be tuned. Hence, for SVR, the parameters to be tuned are C and σ_{kernel} respectively. Similarly, the parameters to be tuned for ReGEPSVR are $\hat{\delta}_1$, $\hat{\delta}_2$ and σ_{kernel} respectively. Further, normalization is applied on the input features of UCI datasets.

For comparing results over UCI datasets, we consider the standard tenfold cross-validation methodology. This yields 10 results corresponding to each fold, of which the mean and variance values are computed and compared. The algorithms on all these datasets are evaluated on the basis of their:

- NMSE: (Normalized Mean Square Error)
  Without loss of generality, let m be the number of testing samples, y_i be the real output value of sample x_i, $\hat{y}_i$ be the predicted value of sample x_i, and $\bar{y} = \frac{1}{m}\sum_i y_i$ be the average value of $y_1, \ldots.y_m$. Then NMSE is defined as

$$NMSE = \frac{(m-1)}{(m)} \times \frac{\sum_{i=1}^{m}(y_i - \hat{y}_i)^2}{\sum_{i=1}^{m}(y_i - \bar{y})^2}. \tag{2.44}$$

A small NMSE value means good agreement between estimations and real-values.

We consider four benchmark datasets, including the Boston Housing, Machine CPU, Servo, and Auto-price datasets, obtained from the UCI repository.

The Boston Housing dataset consists of 506 samples. Each sample has thirteen features which designate the quantities that influence the price of a house in Boston suburb and an output feature which is the house-price in thousands of dollars. The Machine CPU dataset concerns Relative CPU performance data. It consists of 209 cases, with seven continuous features, which are MYCT, MMIN, MMAX, CACH, CHMIN, CHMAX, PRP (output). The Servo dataset consists of 167 samples and covers an extremely nonlinear phenomenon-predicting the rise time of a servo mechanism in terms of two continuous gain settings and two discrete choices of mechanical linkages. The Auto price dataset consists of 159 samples with fifteen features.

Results for comparison between ReGEPSVR and SVR for the four UCI datasets are given in Table 2.1.

Table 2.1 UCI datasets: NMSE comparisons

Dataset	NMSE	NMSE
	TSVR	ReGEPSVR
Boston Housing	0.154 ± 0.053	0.140 ± 0.002
Machine CPU	0.253 ± 0.029	0.198 ± 0.023
Servo	0.185 ± 0.125	0.180 ± 0.027
Auto Price	0.510 ± 0.142	0.223 ± 0.022

2.5 Conclusions

This chapter presents the GEPSVM formulation of Mangasarian and Wild [1] for binary data classification and discusses its advantages over the traditional SVM formulations. We also discuss two variants of the basic GEPSVM formulation which seem to reduce overall computational effort over that of GEPSVM. These are ReGEPSVM formulation of Guarracino et al. [2], and Improved GEPSVM formulation of Shao et al. [3]. In ReGEPSVM we get only a single generalized eigenvalue problem, while Improved GEPSVM deals with two simple eigenvalue problems. Taking motivation from Bi and Bennett [9], a regression analogue of GEPSVM is also discussed. This regression formulation is due to Khemchandani et al. [4] and is termed as Generalized Eigenvalue Proximal Support Vector Regression (GEPSVR). A natural variant of GEPSVR, namely ReGEPSVR is also presented here.

References

1. Mangasarian, O. L., & Wild, E. W. (2006). Multisurface proximal support vector machine classification via generalized eigenvalues. *IEEE Transactions on Pattern Analysis and Machine Intelligence*, *28*(1), 69–74.
2. Guarracino, M. R., Cifarelli, C., Seref, O., & Pardalos, P. M. (2007). A classification method based on generalized eigenvalue problems. *Optimization Methods and Software*, *22*(1), 73–81.
3. Shao, Y.-H., Deng, N.-Y., Chen, W.-J., & Wang, Z. (2013). Improved generalized eigenvalue proximal support vector machine. *IEEE Signal Processing Letters*, *20*(3), 213–216.
4. Khemchandani, R., Karpatne, A., & Chandra, S. (2011). Generalized eigenvalue proximal support vector regressor. *Expert Systems with Applications*, *38*, 13136–13142.
5. Parlett, B. N. (1998). *The symmetric eigenvalue problem: Classics in applied mathematics* (Vol. 20). Philadelphia: SIAM.
6. Tikhonov, A. N., & Arsenin, V. Y. (1977). *Solutions of Ill-posed problems*. New York: Wiley.
7. Saad, Y. (1992). *Numerical methods for large eigenvalue problems*. New York: Halsted Press.
8. Saigal, P., & Khemchandani, R. (2015) Nonparallel hyperplane classifiers for multi-category classification. In *IEEE Workshop on Computational Intelligence: Theories, Applications and Future Directions (WCI)*. Indian Institude of Technology, Kanpur.
9. Bi, J., & Bennett, K. P. (2003). A geometric approach to support vector regression. *Neurocomputing*, *55*, 79–108.
10. Lázarao, M., Santamaría, I., Péreze-Cruz, F., & Artés-Rodríguez, A. (2005). Support vector regression for the simultaneous learning of a multivariate function and its derivative. *Neurocomputing*, *69*, 42–61.
11. Alpaydin, E., & Kaynak, C. (1998). UCI Machine Learning Repository, Irvine, CA: University of California, Department of Information and Computer Sciences. http://archive.ics.uci.edu/ml.
12. Duda, R., Hart, P., & Stork, D. (2001). *Pattern classification*. New York: Wiley.

Chapter 3
Twin Support Vector Machines (TWSVM) for Classification

3.1 Introduction

In Chap. 2, we have presented the formulation of generalized eigenvalue proximal support vector machine (GEPSVM) for binary data classification. This formulation of Mangasarian and Wild [1] aims at generating two non-parallel planes such that each plane is close to one of the two classes and is as far as possible from the other class.

The development of Twin Support Vector Machine (TWSVM) proposed by Jayadeva et al. [2] is greatly motivated by GEPSVM. In the spirit of GEPSVM, TWSVM formulation also aims at generating two non-parallel planes such that each plane is closer to one of the two classes and is as far as possible from the other class. But similarities between TWSVM and GEPSVM ends here. The formulation of TWSVM is totally different from that of GEPSVM. In GEPSVM, the non-parallel planes are determined by evaluating the eigenvectors corresponding to the smallest eigenvalue of two related generalized eigenvalue problems. The TWSVM formulation is very much in line with the standard SVM formulation. However, TWSVM formulation differs from SVM formulation in one fundamental way. In TWSVM, we solve a pair of quadratic programming problems (QPP's), where as in SVM, we solve a single QPP. In SVM formulation, the QPP has all data points in constraints, but in the formulation of TWSVM, they are distributed. Here patterns of one class give the constraints of other QPP and vice versa. This strategy of solving two smaller sized QPP's, rather than a large QPP, makes TWSVM work faster than standard SVM.

This Chapter is organized as follows. Section 3.2 develops the details of linear TWSVM, while the nonlinear kernel version is presented in Sect. 3.3. Section 3.4 deals with the experimental results for linear and nonlinear TWSVM models. Certain advantages and possible drawbacks of TWSVM are discussed in Sect. 3.5. Numerous modifications of the original TWSVM formulation have appeared in the literature and a couple of these, which are directly related with TWSVM, are presented in Sect. 3.6. These are Twin Bounded Support Vector Machine formulation due to Shao et al. [3] and Improved Twin Support Vector Machine formulation due to Tian et al.

Jayadeva et al., *Twin Support Vector Machines*, Studies in Computational Intelligence 659, DOI 10.1007/978-3-319-46186-1_3

[4]. Some other variants of TWSVM are later studied in Chap. 5. Section 3.7 contains certain concluding remarks on TWSVM.

3.2 Linear TWSVM for Binary Data Classification

Let the training set T_C for the given binary data classification problem be

$$T_C = \{(x^{(i)}, y_i), x^{(i)} \in \mathbb{R}^n, y_i \in \{-1, +1\}, (i = 1, 2, ..., m)\}. \tag{3.1}$$

Let there be m_1 patterns in class $+1$ and m_2 patterns in class -1, with $m_1 + m_2 = m$. We form the $(m_1 \times n)$ matrix A with its ith row as $(x^{(i)})^T$ for which $y_i = +1$. The $(m_2 \times n)$ matrix B is constructed similarly with data points $x^{(i)}$ for which $y_i = -1$.

The linear TWSVM formulation consists of determining two hyperplanes $x^T w_1 + b_1 = 0$ and $x^T w_2 + b_2 = 0$ by solving the following pair of quadratic programming problems

$$(TWSVM1) \quad \underset{(w_1,\, b_1,\, q_1)}{Min} \quad \frac{1}{2}\|(Aw_1 + e_1 b_1)\|_2 + C_1 e_2^T q_1$$
$$\text{subject to}$$
$$-(Bw_1 + e_2 b_1) + q_1 \geq e_2,$$
$$q_1 \geq 0, \tag{3.2}$$

and

$$(TWSVM2) \quad \underset{(w_2,\, b_2,\, q_2)}{Min} \quad \frac{1}{2}\|(Bw_2 + e_2 b_2)\|_2 + C_2 e_1^T q_2$$
$$\text{subject to}$$
$$(Aw_2 + e_1 b_2) + q_2 \geq e_1,$$
$$q_2 \geq 0. \tag{3.3}$$

Here $w_1 \in \mathbb{R}^n$, $w_2 \in \mathbb{R}^n$, $b_1 \in \mathbb{R}$, $b_2 \in \mathbb{R}$, $q_1 \in \mathbb{R}^{m_2}$ and $q_2 \in \mathbb{R}^{m_1}$. Also $C_1 > 0$, $C_2 > 0$ are parameters, and $e_1 \in \mathbb{R}^{m_1}$, $e_2 \in \mathbb{R}^{m_2}$ are vectors of 'ones', i.e. each component is 'one' only.

Let us try to interpret the formulation of (TWSVM1) given at (3.2). The first term in the objective function of (TWSVM1) is the sum of squared distances from the hyperplane $x^T w_1 + b_1 = 0$ to points of class $+1$. Therefore, its minimization tends to keep the hyperplane $x^T w_1 + b_1 = 0$ close to the points of class $+1$. The constraints in (3.2) require the hyperplane $x^T w_1 + b_1 = 0$ to be at a distance of at least 1 from points of class -1. A vector of error variables q_1 is used to measure the error wherever the hyperplane is closer than this minimum distance of 1. The second term of the objective function of (TWSVM1) minimizes the sum of error variables, thus attempting to minimize mis-classification due to points belonging to class -1.

The parameters $C_1 > 0$ does the trade-off between the minimization of the two terms in the objective function of (TWSVM1). A similar interpretation may also be given to the formulation (TWSVM2) given at (3.3).

Thus, TWSVM formulation consists of a pair of quadratic programming problems (3.2) and (3.3) such that, in each QPP, the objective function corresponds to a particular class and the constraints are determined by patterns of the other class. As a consequence of this strategy, TWSVM formulation gives rise to two smaller sized QPPs, unlike the standard SVM formulation where a single large QPP is obtained. In (TWSVM1), patterns of class $+1$ are clustered around the plane $x^T w_1 + b_1 = 0$. Similarly, in (TWSVM2), patterns of class -1 cluster around the plane $x^T w_2 + b_2 = 0$.

We observe that TWSVM is approximately four times faster then the usual SVM. This is because the complexity of the usual SVM is no more than m^3, and TWSVM solves two problems, namely (3.2) and (3.3), each of size roughly $(m/2)$. Thus the ratio of run times is approximately $\left[(m^3) / \left(2 \times \left(\frac{m}{2}\right)^3\right)\right] = 4$.

At this stage, we give two simple examples to visually illustrate TWSVM and GEPSVM. Figures 3.1 and 3.2 illustrates the classifier obtained for the two examples by using GEPSVM and TWSVM, respectively. The data consists of points in $\mathbb{R}^2$. Points of class 1 and -1 are denoted by two different shapes. The training set accuracy for TWSVM is 100 percent in both the examples, whereas, for GEPSVM, it is 70 percent and 61.53 percent, respectively.

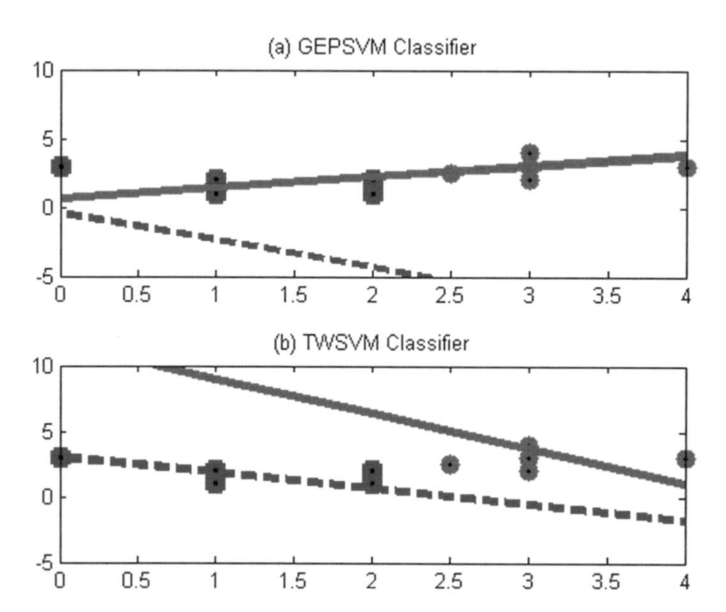

Fig. 3.1 **a** GEPSVM Classifier. **b** TWSVM Classifier. Points of class 1 and -1 are denoted by different shapes

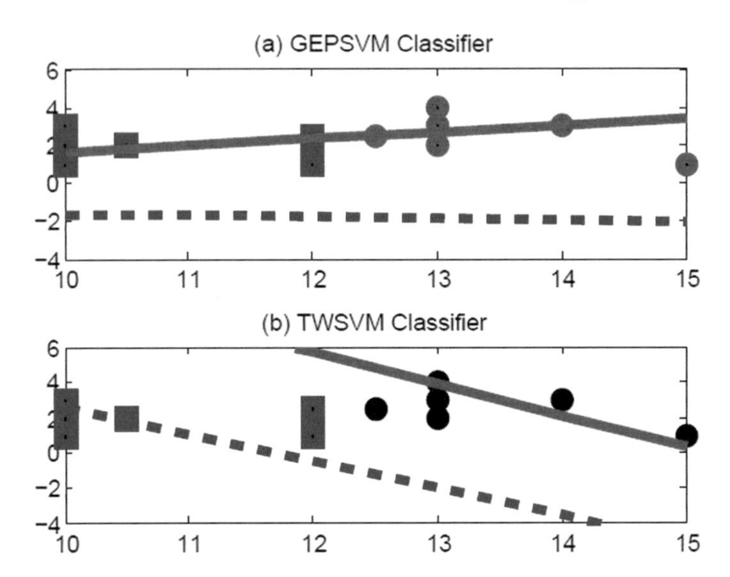

Fig. 3.2 **a** GEPSVM Classifier. **b** TWSVM Classifier. Points of class 1 and -1 are denoted by different shapes

3.2.1 Dual Formulation of Linear TWSVM

Taking motivation from the standard SVM methodology, we derive the dual formulation of TWSVM. Obviously this requires the duals of (TWSVM1) and (TWSVM2). We first consider (TWSVM1) and write its Wolfe dual. For this, we use the Lagrangian corresponding to the problem (TWSVM1) which is given by

$$\begin{aligned} L(w_1, b_1, q_1, \alpha, \beta) = &\frac{1}{2}(Aw_1 + e_1 b_1)^T (Aw_1 + e_1 b_1) + C_1 e_2^T q_1 \\ &- \alpha^T(-(Bw_1 + e_2 b_1) + q_1 - e_2) - \beta^T q_1, \end{aligned} \tag{3.4}$$

where $\alpha = (\alpha_1, \alpha_2 \dots \alpha_{m_2})^T$, and $\beta = (\beta_1, \beta_2 \dots \beta_{m_2})^T$ are the vectors of Lagrange multipliers. As (TWSVM1) is a convex optimization problem, the Karush–Kuhn–Tucker (K. K. T) optimality conditions are both necessary and sufficient (Mangasarian [5], Chandra et al. [6]). Therefore, we write K.K.T conditions for (TWSVM1) and get the following

$$A^T(Aw_1 + e_1 b_1) + B^T \alpha = 0, \tag{3.5}$$

$$e_1^T(Aw_1 + e_1 b_1) + e_2^T \alpha = 0, \tag{3.6}$$

$$C_1 e_2 - \alpha - \beta = 0, \tag{3.7}$$

$$-(Bw_1 + e_2 b_1) + q_1 \geq e_2, \tag{3.8}$$

$$\alpha^T(-(Bw_1 + e_2 b_1) + q_1 - e_2) = 0, \tag{3.9}$$

$$\beta^T q_1 = 0, \tag{3.10}$$

$$\alpha \geq 0, \quad \beta \geq 0, \quad q_1 \geq 0. \tag{3.11}$$

Since $\beta \geq 0$, from (3.7) we have $\alpha \leq C_1$. But $\alpha \geq 0$ and hence $0 \leq \alpha \leq C_1$.
Next, we combine (3.5) and (3.6) and get

$$[A^T \quad e_1^T][A \quad e_1][w_1 \quad b_1]^T + [B^T \quad e_2^T]\alpha = 0. \tag{3.12}$$

We now define

$$H = [A \quad e_1], \quad G = [B \quad e_2], \quad \text{and} \quad u = [w_1, b_1]^T. \tag{3.13}$$

Then (4.28) may be written as

$$(H^T H)u + G^T \alpha = 0. \tag{3.14}$$

If $H^T H$ is invertible, then (3.14) gives

$$u = -(H^T H)^{-1} G^T \alpha. \tag{3.15}$$

Here, we note that $H^T H$ is always positive semidefinite but it does not guarantee that $H^T H$ is invertible. Therefore, on the lines of the regularization term introduced in such as Saunders et al. [7], we introduce a regularization term ϵI, $\epsilon > 0$, I being the identity matrix of appropriate dimension, to take care of problems due to possible ill conditioning of $H^T H$. This makes the matrix $(H^T H + \epsilon I)$ invertible and therefore (3.15) gets modified to

$$u = -(H^T H + \epsilon I)^{-1} G^T \alpha. \tag{3.16}$$

In the following we shall continue to use (3.15) instead of (3.16) with the understanding that, if the need be, (3.16) is to be used for the determination of u.

The Wolfe dual (Mangasarian [5], Chandra et al. [6]) of (TWSVM1) is given by

$$\begin{aligned}
&Max \quad L(w_1, b_1, q_1, \alpha, \beta) \\
&\text{subject to} \\
&\qquad \nabla_{w_1} L(w_1, b_1, q_1, \alpha, \beta) = 0, \\
&\qquad \frac{\partial L}{\partial b_1} = 0, \\
&\qquad \frac{\partial L}{\partial q_1} = 0, \\
&\qquad \alpha \geq 0, \quad \beta \geq 0.
\end{aligned}$$

Now using the K.K.T conditions (3.5)–(3.11), and making use of (3.15), we obtain the Wolfe dual of (TWSVM1) as follows

$$(DTWSVM1) \qquad \underset{\alpha}{Max} \quad e_2^T\alpha - \frac{1}{2}\alpha^T G(H^T H)^{-1} G^T \alpha$$
subject to
$$0 \le \alpha \le C_1.$$

In a similar manner, we consider (TWSVM2) and obtain its dual as

$$(DTWSVM2) \qquad \underset{\nu}{Max} \quad e_1^T\nu - \frac{1}{2}\nu^T J(Q^T Q)^{-1} J^T \nu$$
subject to
$$0 \le \nu \le C_2. \tag{3.17}$$

Here $J = [A \quad e_1]$, $Q = [B \quad e_2]$ and the vector $v = [w_2, b_2]^T$ is given by

$$v = (Q^T Q)^{-1} J^T \nu. \tag{3.18}$$

We solve (DTWSVM1) and (DTWSVM2) to get vectors α and ν respectively. Once α and ν are known, we use (3.15) and (3.18) to get vectors $u = [w_1, b_1]^T$ and $v = [w_2, b_2]^T$ respectively. The vectors u and v so determined give the separating planes

$$x^T w_1 + b_1 = 0 \quad \text{and} \quad x^T w_2 + b_2 = 0. \tag{3.19}$$

A new data point $x \in \mathbb{R}^n$ is assigned to class r $(r = 1, 2)$, depending on which of the two planes given by (3.19) it lies closer to. Thus

$$\text{class}(x) = \underset{i=1,2}{arg\,Min}(d_r(x)),$$

where

$$d_r(x) = \frac{|x^T w_r + b_r|}{\|w^{(r)}\|}. \tag{3.20}$$

Remark 3.2.1 From the Karush–Kuhn–Tucker conditions (3.5)–(3.11), we observe that patterns of class -1 for which $0 < \alpha_i < C_1$, $(i = 1, 2 \ldots, m_2)$, lie on the hyperplane given by $x^T w_1 + b_1 = -1$. Taking motivation from standard SVM terminology, we define such patterns of class -1 as support vectors of class $+1$ with respect to class -1, as they play an important role in determining the required plane. A similar definition of support vectors of class -1 with respect to class $+1$ follows analogously.

3.3 The Nonlinear Kernel TWSVM Classifier

In this section, we extend the results of linear TWSVM to kernel TWSVM. For this, we consider the following kernel generated surfaces instead of planes.

$$K(x^T, C^T)u_1 + b_1 = 0 \quad \text{and} \quad K(x^T, C^T)u_2 + b_2 = 0, \tag{3.21}$$

where $C^T = [A \quad B]^T$, and K is the kernel matrix corresponding to an appropriately chosen kernel function. Also $u_1, u_2 \in \mathbb{R}^m$.

The motivation for choosing the kernel generated surfaces in the form of (3.21) is that by choosing $K(x^T, C^T) = x^T C$ and defining $w_1 = C^T u_1$, the hyperplanes (3.19) corresponding to linear TWSVM are obtained as a special case of (3.21). Therefore, in line with the arguments in Section (3.2), we construct optimization problems (KTWSVM1) and (KTWSVM2) as follows

$$(KTWSVM1) \qquad \underset{(u_1,\, b_1,\, q_1)}{Min} \quad \frac{1}{2}\|(K(A, C^T)u_1 + e_1 b_1)\|^2 + C_1 e_2^T q_1$$
$$\text{subject to}$$
$$-(K(B, C^T)u_1 + e_2 b_1) + q_1 \geq e_2,$$
$$q_1 \geq 0, \tag{3.22}$$

and

$$(KTWSVM2) \qquad \underset{(u_2,\, b_2,\, q_2)}{Min} \quad \frac{1}{2}\|(K(B, C^T)u_2 + e_2 b_2)\|^2 + C_2 e_1^T q_2$$
$$\text{subject to}$$
$$(K(A, C^T)u_2 + e_1 b_2) + q_2 \geq e_1,$$
$$q_2 \geq 0. \tag{3.23}$$

where $C_1 > 0$, $C_2 > 0$ are trade-off parameters.

Next, consider the problem (KTWSVM1) and define its Lagrangian L as follows

$$\begin{aligned} L(u_1, b_1, q_1, \alpha, \beta) = {} & \frac{1}{2}\|(K(A, C^T)u_1 + e_1 b_1)\|^2 - \alpha^T(-(K(B, C^T)u_1 + e_2 b_1 \\ & + q_1 - e_2) + C_1 e_2^T q_1 - \beta^T q_1, \end{aligned} \tag{3.24}$$

where $\alpha = (\alpha_1, \alpha_2 \ldots \alpha_{m_2})^T$, and $\beta = (\beta_1, \beta_2 \ldots \beta_{m_2})^T$ are the vectors of Lagrange multipliers.

We now write K.K.T optimality conditions for (KTWSVM1) and obtain

$$K(A, C^T)^T(K(A, C^T)u_1 + e_1 b_1) + K(B, C^T)^T \alpha = 0, \tag{3.25}$$
$$e_1^T(K(A, C^T)u_1 + e_1 b_1) + e_2^T \alpha = 0, \tag{3.26}$$

$$C_1e_2 - \alpha - \beta = 0, \tag{3.27}$$

$$-(K(B, C^T)u_1 + e_2b_1) + q_1 \geq e_2, \tag{3.28}$$

$$\alpha^T(-(K(B, C^T)u_1 + e_2b_1) + q_1 - e_2) = 0, \tag{3.29}$$

$$\beta^T q_1 = 0, \tag{3.30}$$

$$\alpha \geq 0, \quad \beta \geq 0, \quad q_1 \geq 0. \tag{3.31}$$

From (3.25) and (3.26) and get

$$[K(A, C^T)^T \quad e_1^T][K(A, C^T) \quad e_1][u_1 \quad b_1]^T + [K(B, C^T)^T \quad e_2^T]\alpha = 0. \tag{3.32}$$

Let

$$S = [K(A, C^T) \quad e_1], \quad R = [K(B, C^T) \quad e_2] \quad \text{and} \quad z_1 = [u_1, b_1]^T. \tag{3.33}$$

Then (3.32) can be rewritten as

$$(S^T S)z_1 + R^T\alpha = 0, \quad \text{i.e.} \quad z_1 = -(S^T S)^{-1}R^T\alpha. \tag{3.34}$$

Now using the K.K.T conditions (3.25)–(3.31), and (3.34), we obtain the Wolfe dual (KDTWSVM1) of (KTWSVM1) as follows

$$(KDTWSVM1) \qquad \underset{\alpha}{Max} \quad e_2^T\alpha - \frac{1}{2}\alpha^T R(S^T S)^{-1}R^T\alpha$$

subject to

$$0 \leq \alpha \leq C_1. \tag{3.35}$$

The Wolfe dual (KDTWSVM2) of (KTWSVM2) is obtained analogously as

$$(KDTWSVM2) \qquad \underset{\nu}{Max} \quad e_1^T\nu - \frac{1}{2}\nu^T L(J^T J)^{-1}L^T\nu$$

subject to

$$0 \leq \nu \leq C_2. \tag{3.36}$$

Here, $L = [K(A, C^T) \quad e_1]$ and $J = [K(B, C^T) \quad e_2]$. The augmented vector $z_2 = [u_2, b_2]^T$ is given by $z_2 = (J^T J)^{-1}L^T\nu$, which determines the plane $K(x^T, C^T)u_2 + b_2 = 0$.

Once (KDTWSVM1) and (KDTWSVM2) are solved to obtain the surfaces (3.21), a new point $x \in \mathbb{R}^n$ is assigned to class $+1$ or -1, in a manner similar to the linear case.

3.4 Experimental Results

The Twin Support Vector Machine (TWSVM), GEPSVM and SVM data classification methods were implemented by using MATLAB 7 running on a PC with an Intel P4 processor (3 GHz) with 1 GB RAM. The methods were evaluated on datasets from the UCI Machine Learning Repository. Generalization error was determined by following the standard ten fold cross-validation methodology.

Table 3.1 summarizes TWSVM performance on some benchmark datasets available at the UCI machine learning repository. The table compares the performance of the TWSVM classifier with that of SVM and GEPSVM [1]. Optimal values of c_1 and c_2 were obtained by using a tuning set comprising of 10% of the dataset. Table 3.2 compares the performance of the TWSVM classifier with that of SVM and GEPSVM [1] using a RBF kernel. In case of the RBF kernel, we have employed a rectangular kernel using 80% of the data. Table 3.3 compares the training time for ten folds, of SVM with that of TWSVM. The TWSVM training time has been determined for two cases: the first when an executable file is used, and secondly, when a dynamic linked library (DLL) file is used. The table indicates that TWSVM is not just effective, but also almost four times faster than a conventional SVM, because it solves two quadratic programming problems of a smaller size instead of a single QPP of a very large size.

Table 3.1 Test set accuracy (as percentages) with a linear kernel

Dataset	TWSVM	GEPSVM	SVM
Heart-statlog (270×14)	84.44±4.32	84.81±3.87	84.07±4.40
Heart-c (303×14)	83.80±5.53	84.44±5.27	82.82±5.15
Hepatitis (155×19)	80.79±12.24	58.29±19.07	80.00±8.30
Ionosphere (351×34)	88.03±2.81	75.19±5.50	86.04 ±2.37
Sonar (208×60)	77.26±10.10	66.76±10.75	79.79±5.31
Votes (435×16)	96.08±3.29	91.93±3.18	94.50±2.71
Pima-Indian (768×8)	73.70±3.97	74.60±5.07	76.68±2.90
Australian (690×14)	85.80±5.05	85.65±4.60	85.51±4.58
CMC (1473×9)	67.28±2.21	65.99±2.30	67.82±2.63

Table 3.2 Percentage test set accuracy with a RBF kernel. (* marked) Testing accuracy figures have been obtained from [1]

Dataset	TWSVM	SVM	GEPSVM
Hepatitis	82.67±10.04	83.13±11.25	78.25±11.79
WPBC	81.92±8.98	79.92±9.18	62.7*
BUPA liver	67.83±6.49	58.32±8.20	63.8*
Votes	94.72±4.72	94.94±4.33	94.2*

Table 3.3 Training times (in seconds)

Dataset	TWSVM (EXE file)	TWSVM (DLL file)	SVM (DLL file)
Hepatitis (155×19)	4.37	4.85	12.7
Sonar (208×60)	4.62	6.64	24.9
Heart-statlog (270×14)	4.72	11.3	50.9
Heart-c (303×14)	8.37	14.92	68.2
Ionosphere (351×34)	9.93	25.9	102.2
Votes (435×16)	12.8	45.8	189.4
Australian (690×14)	37.4	142.1	799.2
Pima-Indian (768×8)	56.9	231.5	1078.6
CMC (1473×9)	63.4	1737.9	6827.8

3.5 Certain Advantages and Possible Drawbacks of TWSVM Formulation

The TWSVM formulation has several advantages over the standard SVM type formulation. Some of these are

(i) The dual problems (DTWSVM1) and (DTWSVM2) have m_1 and m_2 variables respectively, as opposed to $m = m_1 + m_2$ variables in the standard SVM. This strategy of solving a pair of smaller sized QPP's instead of a large one, makes the learning speed of TWSVM approximately four times faster than that of SVM.
(ii) TWSVM uses the quadratic loss function and therefore fully considers the prior information within classes in the data. This makes TWSVM less sensitive to the noise.
(iii) TWSVM is useful for automatically discovering two dimensional projection of the data.

But the TWSVM formulation still has some drawbacks which we list below

(i) In the primal problems (TWSVM1) and (TWSVM2), only the empirical risk is minimized, whereas the structural SVM formulation is based on the structural risk minimization principle.
(ii) Though TWSVM solves two smaller sized QPP's, it needs the explicit knowledge of the inverse of matrices $H^T H$ and $Q^T Q$. The requirement of evaluating $(H^T H)^{-1}$ and $(Q^T Q)^{-1}$ explicitly puts severe limitation on the application of TWSVM for very large data sets. Further while evaluating the computational complexity of TWSVM, the cost of computing $(H^T H)^{-1}$ and $(Q^T Q)^{-1}$ should also be included.
(iii) As quadratic loss function is involved in the TWSVM formulation, almost all datapoints are involved in the determination of the final decision function. As a

consequence, TWSVM loses sparseness property which is an important feature of the SVM formulation.

(iv) In the case of standard SVM formulation, the kernel formulation is very natural and direct. Therefore the kernel SVM formulation with the linear kernel is equivalent to the linear SVM. In the case of TWSVM, the nonlinear TWSVM with the linear kernel is not equivalent to the linear TWSVM.

Many researchers have attempted to modify the basic TWSVM formulation so as to take into consideration some of the above drawbacks and we present two of these in the sequel.

3.6 Twin Bounded Support Vector Machine

It is well known that one significant advantage of SVM is the fact that it is based on the structural risk minimization principle, but only the empirical risk is considered in the primal problems (TWSVM1) and (TWSVM2) of TWSVM. Also in the dual formulation of TWSVM, the inverse matrices $(H^T H)^{-1}$ and $(Q^T Q)^{-1}$ appear explicitly. This implies that the TWSVM formulation implicitly assumes that the matrices $(H^T H)$ and $(Q^T Q)$ are non singular. However, this additional requirement can not always be satisfied. This point was certainly noted in Jayadeva et al. [2] where an appropriate regularization technique was suggested to handle this scenario. But then this amounts to solving an approximate problem and not the real TWSVM problem.

Shao et al. [3] in (2011) suggested a variant of TWSVM which improves the original TWSVM formulation. The stated improvements include adherence to the structural risk minimization principle, automatically getting matrices for the dual formulation whose inverses are guaranteed and can be used for successive overrelaxation methodology (SOR) methodology. Shao et al. [3] termed their model as Twin Bounded Support Vector Machine (TBSVM) which we discuss in the next sub-section.

3.6.1 Linear TBSVM

The linear TBSVM solves the following two primal problems

$$(TBSVM1) \qquad \underset{(w_1,\, b_1,\, q_1)}{Min} \quad \frac{1}{2}((Aw_1 + e_1 b_1))^T((Aw_1 + e_1 b_1)) + \frac{1}{2}C_3(\|w_1\|^2 + (b_1)^2) + C_1 e_2^T q_1$$

subject to

$$\begin{aligned} -(Bw_1 + e_2 b_1) + q_1 &\geq e_2, \\ q_1 &\geq 0, \end{aligned} \tag{3.37}$$

and

$$(TBSVM2) \qquad \underset{(w_2,\, b_2,\, q_2)}{Min} \quad \frac{1}{2}((Bw_2 + e_2b_2))^T((Bw_2 + e_2b_2)) + \frac{1}{2}C_4(\|w_2\|^2 + (b_2)^2) + C_2e_1^Tq_2$$

subject to

$$(Aw_2 + e_1b_2) + q_2 \geq e_1,$$
$$q_2 \geq 0. \tag{3.38}$$

where C_i, $(i = 1, 2, 3, 4)$ are penalty parameters, and other notations are same as in (3.2) and (3.3).

Looking at the primal problems (TWSVM1)-(TWSVM2) and (TBSVM1)-(TBSVM2), we notice that (TBSVM1) has an extra term $\frac{1}{2}C_3(\|w_1\|^2 + (b_1)^2)$ while (TBSVM2) has an extra term $\frac{1}{2}C_4(\|w_2\|^2 + (b_2)^2)$. Let us consider the effect of these extra terms and for this let us consider (TBSVM1) only, the arguments for (TBSVM2) are analogous.

In the following section authors in [3] argue that due to the extra term $\frac{1}{2}C_3(\|w_1\|^2 + (b_1)^2)$, the structural risk is minimized in (3.37). Let us recollect that in SVM, the structural risk minimization is implemented by maximizing the margin which is measured by the Euclidean distance between the two supporting hyperplanes. In the context of TWSVM, the corresponding margin between the two classes can be measured by the distance between the proximal hyperplane $x^Tw_1 + b_1 = 0$ and the bounding hyperplane $x^Tw_1 + b_1 = -1$. This distance is $1/\|w_1\|^2$, which is called the one sided margin between two classes with respect to the hyperplane $x^Tw_1 + b_1 = 0$. This suggests for the inclusion of the term $\|w_1\|^2$ in the objective function of (3.37). We also have an extra term $(b_1)^2$ which is motivated by Fung and Mangasarian [8] in the context of Proximal Support Vector Machine (PSVM) formulation. In fact, if we define $X = [x^T, 1]^T$, $W_1 = [w_1, b_1]$ then the proximal plane in $\mathbb{R}^{n+1}$ is $X^TW_1 = 0$. Therefore, the required distance is $1/\|W_1\|$, i.e. $1/\sqrt{(\|w_1\|^2 + (b_1)^2)}$. Similar arguments give justification for the inclusion of the term $\|w_2\|^2 + (b_2)^2$ in the objective function of problem (3.38).

Though Shao et al. [3] claimed that the inclusion of additional terms $C_3/(\|w_1\|^2 + b_1^2)$ in (TBSVM1), and $C_4/(\|w_2\|^2 + b_2^2)$ in (TBSVM2) does the structural risk minimization, their arguments are only motivational which attempt to justify their claims. Unlike SVM, a proper mathematical theory of TBSVM based on statistical learning theory (SLT) is still not available and this aspect needs to be further explored.

In order to get the solutions of problems (TBSVM1) and (TBSVM2) we need to derive their dual problems. To get the dual (DTBSVM1) of (TBSVM1), we construct the Lagrangian and then follow the standard methodology. For (TBSVM1), the Lagrangian is

$$L(w_1, b_1, q_1, \alpha, \beta) = \frac{1}{2}C_3(\|w_1\|^2 + (b_1)^2) + \frac{1}{2}\|Aw_1 + e_1b_1\|^2 + C_1e_2^Tq_1 - \alpha^T(-(Bw_1 + e_2b_1) + q_1 - e_2) - \beta^Tq_1, \tag{3.39}$$

where $\alpha = (\alpha_1, \alpha_2 \ldots \alpha_{m_2})^T$, and $\beta = (\beta_1, \beta_2 \ldots \beta_{m_2})^T$ are the vectors of Lagrange multipliers.

Now writing the Wolfe dual of (TBSVM1), we get (DTBSVM1) as follows

$$
\begin{aligned}
(DTBSVM1) \qquad & \underset{\alpha}{Max} \quad e_2^T\alpha - \frac{1}{2}\alpha^T G(H^T H + C_3 I)^{-1} G^T \alpha \\
& \text{subject to} \\
& 0 \le \alpha \le C_1.
\end{aligned} \tag{3.40}
$$

In a similar manner, we get the Wolfe dual (DTBSVM2) of (TBSVM2) as

$$
\begin{aligned}
(DTBSVM2) \qquad & \underset{\nu}{Max} \quad e_1^T\nu - \frac{1}{2}\nu^T J(Q^T Q + C_4 I)^{-1} J^T \nu \\
& \text{subject to} \\
& 0 \le \nu \le C_2.
\end{aligned} \tag{3.41}
$$

Here G, H, J and Q are same as in problems (DTWSVM1) and (DTWSVM2).

In problems (DTBSVM1) and (DTBSVM2), the matrices $(H^T H + C_3 I)$ and $(Q^T Q + C_4 I)$ are non singular naturally. Therefore, the inverses $(H^T H + C_3 I)^{-1}$ and $(Q^T Q + C_4 I)^{-1}$ are guaranteed to exist without any extra assumption or modification. The rest of details are similar to that of TWSVM.

3.6.2 Nonlinear TBSVM

As in TWSVM, to extend our results to nonlinear case, we consider the following kernel generated surfaces instead of hyperplanes

$$
K(x^T, C^T)u_1 + b_1 = 0 \quad \text{and} \quad K(x^T, C^T)u_2 + b_2 = 0, \tag{3.42}
$$

where $C^T = [A \quad B]^T$, and K is the kernel matrix corresponding to an appropriately chosen kernel function. For the surface $K(x^T, C^T)u_1 + b_1 = 0$, we construct the optimization problem as

$$
\begin{aligned}
(KTBSVM1) \qquad & \underset{(u_1,\, b_1,\, q_1)}{Min} \quad \frac{1}{2}\|(K(A, C^T)u_1 + e_1 b_1)\|^2 \\
& \qquad + \frac{1}{2}C_3(\|u_1\|^2 + b_1^2) + C_1 e_2^T q_1 \\
& \text{subject to} \\
& -(K(B, C^T)u_1 + e_2 b_1) + q_1 \ge e_2, \\
& q_1 \ge 0,
\end{aligned} \tag{3.43}
$$

where $C_1 > 0$ and $C_3 > 0$ are trade-off parameters.

The dual of (KTBSVM1) is obtained as

$$(DKTBSVM1) \quad \underset{\alpha}{Max} \quad e_2^T\alpha - \frac{1}{2}\alpha^T R(S^T S + C_3 I)^{-1} R^T \alpha$$
$$\text{subject to}$$
$$0 \leq \alpha \leq C_1. \tag{3.44}$$

In a similar manner, the problem (KTBSVM2) is constructed and its dual is obtained as

$$(DKTBSVM2) \quad \underset{\nu}{Max} \quad e_1^T\nu - \frac{1}{2}\nu^T L(J^T J + C_4 I)^{-1} L^T \nu$$
$$\text{subject to}$$
$$0 \leq \nu \leq C_2. \tag{3.45}$$

Here the matrices R, S, L and J are same as in problems (3.35) and (3.36). The rest of details are analogous to that of nonlinear TWSVM.

Remark 3.6.1 Shao et al. [3] implemented their linear TBSVM and nonlinear TBSVM models on various artificial and real life datasets. The experimental results show the effectiveness of these models in both computation time and classification accuracy. In fact Shao et al. [3] used successive overrelaxation technique (SOR) to speed up the training procedure, but explicit inverse of relevant matrices appearing in the duals is still required.

3.7 Improved Twin Support Vector Machine

In a recent work, Tian et al. [4] presented an improved model of twin methodology, termed as Improved Twin Support Vector Machine (ITSVM), for binary data classification. Surprisingly ITSVM is exactly same as TBSVM (Shao et al. [3]) but represented differently. This leads to different Lagrangian function for primal problems in TWSVM and TBSVM, and therefore different dual formulations. It has been shown in Tian et al. [4] that ITSVM does not need to compute the large inverse matrices before training and the kernel trick can be applied directly to ITSVM for the nonlinear case. Further, ITSVM can be solved efficiently by the successive overrelaxation (SOR) and sequential minimization optimization (SMO) techniques, which makes it more suitable for large datasets.

3.7.1 Linear ITSVM

Consider the primal problems (TBSVM1) and (TBSVM2) given at (3.37)–(3.38). Let $Aw_1 + e_1b_1 = p$ and $Bw_2 + e_2b_2 = q$. Then problems (3.37) and (3.38) can be rewritten as

$$(ITSVM1) \qquad \underset{(w_1,\, b_1,\, p,\, q_1)}{Min} \quad \frac{1}{2}p^Tp + \frac{1}{2}C_3(\|w_1\|^2 + (b_1)^2) + C_1e_2^Tq_1$$

subject to

$$\begin{aligned} Aw_1 + e_1b_1 &= p, \\ -(Bw_1 + e_2b_1) + q_1 &\geq e_2, \\ q_1 &\geq 0, \end{aligned} \tag{3.46}$$

and

$$(ITSVM2) \qquad \underset{(w_2,\, b_2,\, q,\, q_2)}{Min} \quad \frac{1}{2}q^Tq + \frac{1}{2}C_4(\|w_2\|^2 + (b_2)^2) + C_2e_1^Tq_2$$

subject to

$$\begin{aligned} Bw_2 + e_2b_2 &= q, \\ (Aw_2 + e_1b_2) + q_2 &\geq e_1, \\ q_2 &\geq 0. \end{aligned} \tag{3.47}$$

We next consider (ITSVM1) and introduce the Lagrangian

$$\begin{aligned} L(w_1, b_1, p, q_1, \alpha, \beta, \lambda) = &\frac{1}{2}C_3(\|w_1\|^2 + (b_1)^2) + \frac{1}{2}p^Tp + C_1e_2^Tq_1 \\ &+ \lambda^T(Aw_1 + e_1b_1 - p) - \alpha^T(-(Bw_1 + e_2b_1) + q_1 - e_2) - \beta^Tq_1, \end{aligned} \tag{3.48}$$

where $\alpha = (\alpha_1, \alpha_2 \ldots \alpha_{m_2})^T, \beta = (\beta_1, \beta_2 \ldots \beta_{m_2})^T$ and $\lambda = (\lambda_1, \lambda_2 \ldots \lambda_{m_1})^T$ are the vectors of Lagrange multipliers. The K.K.T necessary and sufficient optimality conditions for (ITSVM1) are given by

$$C_3w_1 + A^T\lambda + B^T\alpha = 0, \tag{3.49}$$

$$C_3b_1 + e_1^T\lambda + e_2^T\alpha = 0, \tag{3.50}$$

$$\lambda - p = 0, \tag{3.51}$$

$$C_1e_2 - \alpha - \beta = 0, \tag{3.52}$$

$$Aw_1 + e_1b_1 = p, \tag{3.53}$$

$$-(Bw_1 + e_2b_1) + q_1 \geq e_2, \tag{3.54}$$

$$\alpha^T(Bw_1 + e_2b_1 - q_1 + e_2) = 0, \tag{3.55}$$

$$\beta^Tq_1 = 0, \tag{3.56}$$

$$\alpha \geq 0, \quad \beta \geq 0, \quad \lambda \geq 0, \quad q_1 \geq 0. \tag{3.57}$$

Since $\beta \geq 0$, (3.52) gives $0 \leq \alpha \leq C_1 e_2$. Also (3.49) and (3.50) give

$$w_1 = -\frac{1}{C_3}(A^T \lambda + B^T \alpha), \tag{3.58}$$

$$b_1 = -\frac{1}{C_3}(e_1^T \lambda + e_2^T \alpha). \tag{3.59}$$

Now using (3.58), (3.59) and (3.51), we obtain the Wolfe dual of (ITSVM1) as

$$(DITSVM1) \qquad \underset{\lambda, \alpha}{Max} \quad -\frac{1}{2}(\lambda^T, \alpha^T) Q_1 (\lambda^T, \alpha^T)^T + C_3 e_2^T \alpha$$
subject to
$$0 \leq \alpha \leq C_1 e_2. \tag{3.60}$$

Here

$$Q_1 = \begin{bmatrix} AA^T + C_3 I & AB^T \\ BA^T & BB^T \end{bmatrix} + E, \tag{3.61}$$

and I is the $(m_1 \times m_1)$ identity matrix. Also E is the $(m \times m)$ matrix having all entries as 'one'.

Similarly the dual of (ITSVM2) is obtained as

$$(DITSVM2) \qquad \underset{\theta, \nu}{Max} \quad -\frac{1}{2}(\theta^T, \nu^T) Q_2 (\theta^T, \nu^T)^T + C_4 e_1^T \nu$$
subject to
$$0 \leq \nu \leq C_2 e_1, \tag{3.62}$$

where

$$Q_2 = \begin{bmatrix} BB^T + C_4 I & BA^T \\ AB^T & AA^T \end{bmatrix} + E. \tag{3.63}$$

Let (λ^*, α^*) be optimal to (DITSVM1) and (θ^*, ν^*) be optimal to (DITSVM2). Then the required non-parallel hyper planes are $x^T \overline{w}_1 + \overline{b}_1 = 0$ and $x^T \overline{w}_2 + \overline{b}_2 = 0$, where

$$\overline{w}_1 = -\frac{1}{C_3}(A^T \lambda^* + B^T \alpha^*),$$

$$\overline{w}_2 = -\frac{1}{C_4}(B^T \theta^* + A^T \nu^*),$$

$$\bar{b}_1 = -\frac{1}{C_3}(e_1^T\lambda^* + e_2^T\alpha^*),$$

and

$$\bar{b}_2 = -\frac{1}{C_4}(e_2^T\theta^* + e_1^T\nu^*).$$

The linear ITSVM is equivalent to linear TBSVM because the primal problems are same; only they have different representation.

The important point to note here is that problems (DITSVM1) and (DITSVM2) are quadratic programming problems which do not require computation of matrices inverse. In this respect, the ITSVM formulation is certainly attractive in comparison to TWSVM and TBSVM formulations. However major disadvantage of ITSVM is that the matrices Q_1 and Q_2 involve all the patterns of class A and B. In the recent work, Peng et al. [9] presented a L_1 norm version of ITSVM which again needs to optimize a pair of larger sized dual QPPs than TWSVM.

3.7.2 Nonlinear ITSVM

We now present the nonlinear ITSVM formulation. Here unlike the nonlinear TBSVM or TWSVM formulations, we do not need to consider the kernel generated surfaces and construct two new primal problems corresponding to these surfaces. But rather we can introduce the kernel function directly into the problems (3.60) and (3.62) in the same manner as in the case of standard SVM formulation. As a consequence of this construction, it follows that similar to SVM, linear ITSVM is a special case of nonlinear ITSVM with its specific choice of the kernel as the linear kernel. Unfortunately this natural property is not shared by TWSVM or TBSVM formulations.

Let us now introduce the kernel function $K(x, x') =< \phi(x), \phi(x') >$ and the corresponding transformation $z = \phi(x)$ where $z \in \mathcal{H}$, $\mathcal{H}$ being an appropriate Hilbert space, termed as feature space. Therefore, the corresponding primal ITSVM in the feature space are

$$\underset{(w_1,\, b_1,\, p,\, q_1)}{Min} \quad \frac{1}{2}p^Tp + \frac{1}{2}C_3(\|w_1\|^2 + (b_1)^2) + C_1e_2^Tq_1$$
subject to

$$\begin{aligned} \phi(A)w_1 + e_1b_1 &= p, \\ -(\phi(B)w_1 + e_2b_1) + q_1 &\geq e_2, \\ q_1 &\geq 0, \end{aligned} \tag{3.64}$$

and

$$\begin{aligned} \underset{(w_2,\, b_2,\, q,\, q_2)}{Min} \quad & \frac{1}{2}q^T q + \frac{1}{2}C_4(\|w_2\|^2 + (b_2)^2) + C_2 e_1^T q_2 \\ \text{subject to} & \\ & \phi(B)w_2 + e_2 b_2 = q, \\ & (\phi(A)w_2 + e_1 b_2) + q_2 \geq e_1, \\ & q_2 \geq 0. \end{aligned} \tag{3.65}$$

We next consider problem (3.64) and write its Wolfe dual to get

$$\begin{aligned} (DKITSVM1) \quad & \underset{\lambda,\alpha}{\text{Max}} \quad -\tfrac{1}{2}(\lambda^T, \alpha^T)Q_3(\lambda^T, \alpha^T)^T + C_3 e_2^T \alpha \\ & \text{subject to} \\ & 0 \leq \alpha \leq C_1 e_2. \end{aligned} \tag{3.66}$$

Here

$$Q_3 = \begin{bmatrix} K(A^T, A^T) + C_3 I & K(A^T, B^T) \\ K(A^T, B^T) & K(B^T, B^T) \end{bmatrix} + E. \tag{3.67}$$

Similarly the dual (DKITSVM2) is

$$\begin{aligned} (DKITSVM2) \quad & \underset{\theta,\nu}{Max} \quad -\frac{1}{2}(\theta^T, \nu^T)Q_4(\theta^T, \nu^T)^T + C_4 e_1^T \nu \\ & \text{subject to} \\ & \leq \nu \leq C_2 e_1, \end{aligned} \tag{3.68}$$

where

$$Q_4 = \begin{bmatrix} K(B^T, B^T) + C_3 I & K(B^T, B^T) \\ K(B^T, B^T) & K(A^T, A^T) \end{bmatrix} + E, \tag{3.69}$$

Let (λ^*, α^*) and (θ^*, ν^*) be optimal to problems (3.66) and (3.68) respectively. Then the required decision surfaces are

$$K(x^T, A^T)\lambda^* + K(x^T, B^T)\alpha^* + b_1^* = 0,$$
$$K(x^T, B^T)\theta^* + K(x^T, A^T)\nu^* + b_2^* = 0,$$

where

$$b_1^* = e_1^T \lambda^* + e_2^T \alpha^*,$$
$$b_2^* = e_2^T \theta^* + e_1^T \nu^*.$$

Here again we note that in problems (3.66) and (3.68) we do not require the computation of inverse matrices any more. Also these problems degenerate to problems (3.60) and (3.62) corresponding to linear ITSVM when the chosen kernel K is the linear kernel.

Remark 3.7.1 Tian et al. [4] presented two fast solvers to solve various optimization problems involved in ITSVM, namely SOR and SMO type algorithms. This makes ITSVM more suitable to large scale problems. Tian et al. [4] also performed a very detailed numerical experimentation with ITSVM, both on small datasets and very large datasets. The small datasets were the usual publicly available benchmark datasets, while the large datasets were NDC-10k, NDC-50k and NDC-1m using Musicant's NDC data generator [10]. ITSVM performed universally better than TWSVM and TBSVM on most of datasets. On large datasets NDC-10k, NDC-50k and NDC-1m, TWSVM and TBSVM failed because experiments ran out of memory whereas ITSVM produced the classifier.

3.8 Conclusions

This chapter presents TWSVM formulation for binary data classification and discusses its advantages and possible drawbacks over the standard SVM formulation. Though there are several variants of the basic TWSVM formulation available in the literature, we discuss two of these in this chapter. These are TBSVM formulation due to Shao et al. [3] and ITSVM formulation due to Tian et al. [4]. The construction of TBSVM attempts to include structural risk minimization in its formulation and therefore claims its theoretical superiority over TWSVM. The construction of ITSVM is very similar to TBSVM but has different mathematical representation. One added advantage of ITSVM is that it does not require computation of inverses and therefore is more suitable to large datasets. We shall discuss some other relevant variants of TWSVM later in Chap. 5.

References

1. Mangasarian, O. L., & Wild, E. W. (2006). Multisurface proximal support vector machine classification via generalized eigenvalues. *IEEE Transactions on Pattern Analysis and Machine Intelligence*, *28*(1), 69–74.
2. Jayadeva, Khemchandani, R., & Chandra, S. (2007). Twin support vector machines for pattern classification. *IEEE Transactions on Pattern Analysis and Machine Intelligence*, *29*(5), 905–910.
3. Shao, Y.-H., Zhang, C.-H., Wang, X.-B., & Deng, N.-Y. (2011). Improvements on twin support vector machines. *IEEE Transactions on Neural Networks*, *22*(6), 962–968.
4. Tian, Y. J., Ju, X. C., Qi, Z. Q., & Shi, Y. (2013). Improved twin support vector machine. *Science China Mathematics*, *57*(2), 417–432.
5. Mangasarian, O. L. (1994). *Nonlinear programming*. Philadelphia: SIAM.

6. Chandra, S., Jayadeva, & Mehra, A. (2009). *Numerical optimization with applications*. New Delhi: Narosa Publishing House.
7. Saunders, C., Gammerman, A., & Vovk, V. (1998). Ridge regression learning algorithm in dual variables. In *Proceedings of the Fifteenth International Conference on Machine Learning* (pp. 515–521).
8. Fung, G., & Mangasarian, O. L. (2001). Proximal support vector machine classifiers. In F. Provost & R. Srikant (Eds.), *Proceedings of Seventh International Conference on Knowledge Discovery and Data Mining* (pp. 77–86).
9. Peng, X. J., Xu, D., Kong, L., & Chen, D. (2016). L1-norm loss based twin support vector machine for data recognition. *Information Sciences*, *340–341*, 86–103.
10. Musicant, D. R. (1998). NDC: Normally Distributed Clustered Datasets, Computer Sciences Department, University of Wisconsin, Madison.

Chapter 4
TWSVR: Twin Support Vector Machine Based Regression

4.1 Introduction

SVR (Support Vector Regression) is a SVM based approach to study the regression problem. The standard SVR model sets an epsilon tube around data points within which errors are discarded using an epsilon insensitive loss function. We have already presented the standard SVR formulation in Chap. 1.

One of the major theoretical developments in the context of SVR is due to Bi and Bennett [1]. They (Bi and Bennett [1]) presented an intuitive geometric framework for SVR which shows that SVR can be related to SVM for an appropriately constructed classification problem. This result of Bi and Bennett [1] is conceptually very significant, because it suggests that any variant of SVM has possibility of having an analogous SVR formulation. This has been the motivation of introducing GEPSVR in Chap. 2.

In Chap. 3, we have presented the development of TWSVM (Jayadeva et al. [2]) for the binary data classification problem. Since TWSVM has proven its advantage over the standard SVM, it makes sense to look into the possibility of obtaining a regression analogue of TWSVM. In the literature, Peng [3] is credited to initiate the study of regression problem in the twin framework. The work of Peng [3] motivated many researchers to further study regression problem in twin setting, e.g. Xu and Wang [4], Shao et al. [5], Chen et al. [6, 7], Zhao et al. [8], Zhang et al. [9], Peng [10, 11], Balasundaram and Tanveer [12] and Singh et al. [13].

Recently, Khemchandani et al. [14, 15] presented a new framework of Twin Support Vector model to regression problem, termed as TWSVR. Unlike Peng [3], TWSVR is truely inspired by TWSVM, where the upper bound regressor (respectively lower bound regressor) problem deals with the proximity of points in upper tube (respectively lower tube) and at the same time, at least ϵ distance from the points of lower tube (respectively upper tube).

In our presentation here, we shall differentiate between two notations TSVR and TWSVR. TSVR refers to Peng's formulation [3] where as TWSVR refers to

Jayadeva et al., *Twin Support Vector Machines*, Studies in Computational Intelligence 659, DOI 10.1007/978-3-319-46186-1_4

Khemchandani et al. formulation [14, 15]. The terminology of TWSVR seems to be more natural because the basic twin classification formulation is termed as TWSVM.

In this chapter, we also discuss the problem of simultaneous learning of function and its derivative in the context of TWSVR and GEPSVR.

This chapter consists of five main sections, namely, TSVR: Peng's Model, SVR via SVM, TWSVR via TWSVM, A Dual Formulation of TWSVR and simultaneous learning of function and its derivative. Our presentation here is based on Bi and Bennett [1], Peng [3], Goyal [16], Khemchandani et al. [14, 15] and Deng et al. [17].

4.2 TSVR: Peng's Model

Let the training set for the regression problem be given by

$$T_R = \{(x^{(i)}, y_i),\ i = 1, 2 \ldots, l\}, \tag{4.1}$$

where $x^{(i)} \in \mathbb{R}^n$ and $y_i \in \mathbb{R}$.

Following Bi and Bennett [1], we consider the associated classification problem with the training set T_C in $\mathbb{R}^{n+1}$ given by

$$T_C = \{((x^{(i)}, y_i + \epsilon), +1), ((x^{(i)}, y_i - \epsilon), -1), (i = 1, 2 \ldots, l)\}. \tag{4.2}$$

Let A be an $(l \times n)$ matrix whose i^{th} row is the vector $(x^{(i)})^T$. Let $Y = (y_1, y_2 \ldots, y_l)$, $(Y + \epsilon e) = (y_1 + \epsilon, y_2 + \epsilon \ldots, y_l + \epsilon)$ and $(Y - \epsilon e) = (y_1 - \epsilon, y_2 - \epsilon \ldots, y_l - \epsilon)$. Then $f(x) = x^T w + b$ may be identified as a hyperplane in $\mathbb{R}^{n+1}$.

The TSVR formulation as given by Peng [3] consists of following two QPP's

$$\text{(TSVR1)} \qquad \underset{(w_1,\, b_1,\, \xi_1)}{\text{Min}} \quad \frac{1}{2}\|(Y - e\epsilon_1 - (Aw_1 + eb_1))\|_2 + C_1 e^T \xi_1$$

subject to

$$\begin{aligned} Y - e\epsilon_1 - (Aw_1 + eb_1) &\geq -\xi_1, \\ \xi_1 &\geq 0, \end{aligned} \tag{4.3}$$

and

$$\text{(TSVR2)} \qquad \underset{(w_2,\, b_2,\, \xi_2)}{\text{Min}} \quad \frac{1}{2}\|(Y + e\epsilon_2 - (Aw_2 + eb_2))\|_2 + C_2 e^T \xi_2$$

subject to

$$\begin{aligned} (Aw_2 + eb_2) - (Y + e\epsilon_2) &\geq -\xi_2, \\ \xi_2 &\geq 0, \end{aligned} \tag{4.4}$$

where $C_1, C_2 > 0, \epsilon_1, \epsilon_2 > 0$ are parameters, ξ_1, ξ_2 are slack vectors, e denotes vector of ones of appropriate dimension and $\|.\|_2$ denotes the L_2 norm.

Each of the above two QPP is smaller than the one obtained in the classical SVR formulation. Also (TSVR1) finds $f_1(x) = x^T w_1 + b_1$ the down bound regressor and (TSVR2) finds the up bound regressor $f_2(x) = x^T w_2 + b_2$. The final regressor is taken as the mean of up and down bound regressor.

We would now like to make certain remarks on Peng's formulation of TSVR. These remarks not only convince that Peng's formulation is not in the true spirit of TWSVM but also motivate the proposed formulation of Khemchandani et al. [14, 15]. In this context, we have the following lemma.

Lemma 4.2.1 *For the given dataset, let $f(x)$ be the final regressor obtained from (TSVR1) and (TSVR2) when $\epsilon_1 = \epsilon_2 = 0$, and $g(x)$ be the final regressor obtained for any constant value of ϵ_1 and ϵ_2. Then*

$$g(x) = f(x) - (\epsilon_1 - \epsilon_2)/2.$$

Proof Let (w_1, b_1) and (w_2, b_2) be the solutions to (TSVR1) and (TSVR2) respectively for constant ϵ_1 and ϵ_2 so that $g(x) = (x^T w_1 + x^T w_2)/2 + (b_1 + b_2)/2$. Now applying the transformation $(b_1)_{new} = b_1 + \epsilon_1$ to (TSVR1) and $(b_2)_{new} = b_2 - \epsilon_2$ to (TSVR2), we note that the resulting formulations have no ϵ term in it. The final regressor obtained from these transformed formulations will be $f(x)$. It follows from the transformation that $(w_1, b_1 + \epsilon_1)$ and $(w_2, b_2 - \epsilon_2)$ will be the solutions to transformed QPPs. Hence $f(x) = (x^T w_1 + x^T w_2)/2 + (b_1 + b_2)/2 + (\epsilon_1 - \epsilon_2)/2$, thus proving the result.

The understanding to Lemma 4.2.1 comes from analyzing the physical interpretation of the TSVR model. For both the formulations of TSVR, the objective function and the constraints are shifted by the epsilon values in the same direction. This shifting by epsilon in the same direction makes the role of epsilon limited to only shifting the final regressor linearly and not playing any role in the orientation of the final regressor. Lemma 4.2.1 essentially tells that the general regressor $g(x)$ can be obtained by first solving two QPP's (TSVR1) and (TSVR2) by neglecting ϵ_1 and ϵ_2, and then shifting the same by $(\epsilon_1 - \epsilon_2)/2$.

The following remarks are evident from Lemma 4.2.1.

(i) The values of ϵ_1 and ϵ_2 only contribute to linear shift of the final regressor. The orientation of the regressor is independent of values of ϵ_1 and ϵ_2. Therefore in the final hyperplane $y = w^T x + b$, only b is a function of ϵ_1 and ϵ_2 whereas w is independent of ϵ_1 and ϵ_2. This suggest that TSVR formulation is not even in line with the classical Support Vector Regression.

(ii) The regressor is dependent only on one value $\epsilon_1 - \epsilon_2$. This unnecessarily increases the burden of parameter selection of two parameters (ϵ_1, ϵ_2) when in reality, the final regressor depends on only one parameter $(\epsilon_1 - \epsilon_2)$.

(iii) The regressor is independent of values of ϵ_1 and ϵ_2 for $\epsilon_1 = \epsilon_2$. We get the same final regressor, say if $\epsilon_1 = \epsilon_2 = 0.1$ or when $\epsilon_1 = \epsilon_2 = 100$. The Experiment

Section of Peng [3] only shows results for $\epsilon_1 = \epsilon_2$ case, which we have shown is same as not considering $\epsilon's$ at all.

In order to support the above facts empirically, we would provide certain plots in Sect. 4.5

Apart from above points, there is another problem with formulation of Peng [3]. Let us consider (TSVR1) first. For any sample point $(x^{(i)}, y_i)$ and the fitted function $f_1(x)$, if $(y_i - f_1(x^{(i)}) - \epsilon) \geq 0$, then the penalty term is $\frac{1}{2}(y_i - f_1(x^{(i)}) - \epsilon)^2$, and if $(y_i - f_1(x^{(i)}) - \epsilon) \leq 0$, then the penalty term is $\frac{1}{2}(y_i - f_1(x^{(i)}) - \epsilon)^2 + C_1 \mid y_i - f_1(x^{(i)}) - \epsilon \mid$. Thus the penalty term is asymmetric as it gives more penalty on negative deviation of $y_i - f_1(x^{(i)}) - \epsilon$. This is happening because the points of the same class $\left\{(x^{(i)}, y_i - \epsilon), i = 1, \ldots, l\right\}$ are appearing both in the objective function and the constraints, where $\left\{(x^{(i)}, y_i - \epsilon), i = 1, \ldots, l\right\}$ and $\left\{(x^{(i)}, y_i + \epsilon), i = 1, \ldots, l\right\}$ are the two classes of points. Similar arguments hold for (TSVR2) as well. This logic is not consistent with the basic principle of twin methodology.

4.3 SVR via SVM

In this section, we present Bi and Bennett [1] result on ϵ-insensitive regression problem so as to show the equivalence between a given regression problem and an appropriately constructed classification problem. For the purpose of illustrating this equivalence, Bi and Bennett [1] considered a linear regression problem in $\mathbb{R}$ as in Fig. 4.1(a). Here the training set is $\{(x^{(i)}, y_i), i = 1, 2 \ldots, 9\}$. These points are moved up and down by ϵ' in y plane to obtain the points $\{(x^{(i)}, y_i + \epsilon'), i = 1, 2, \ldots, 9\}$ and $\{(x^{(i)}, y_i - \epsilon'), i = 1, 2 \ldots, 9\}$ as shown in Fig. 4.1(b). Now joining the points $(x^{(i)}, y_i + \epsilon')$ and $(x^{(i)}, y_i - \epsilon')$, line segments are obtained. It is easy to verify that any line that passes through these line segments is an ϵ'-insensitive hyperplane. This implies that line separating/classifying the two class points $\{(x^{(i)}, y_i + \epsilon'), i =$

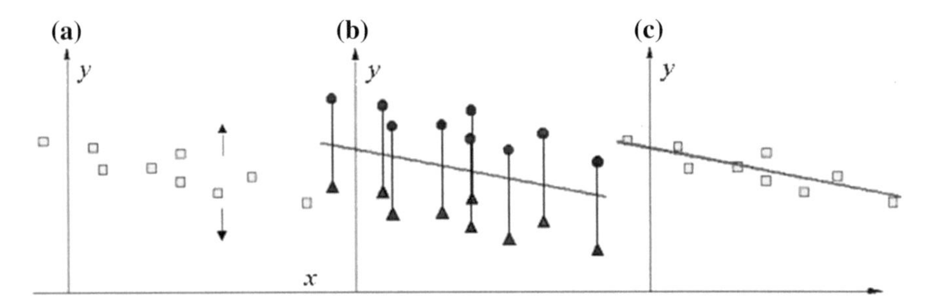

Fig. 4.1 ϵ Band Regression; **a** original data (A_i); **b** shifted data $(A_i^+$ and $A_i^-)$ and separating plane; **c** regression plane

$1, 2, \ldots, 9\}$ and $\{(x^{(i)}, y_i - \epsilon'), i = 1, 2, \ldots, 9\}$ correctly is the required regressor function (Fig. 4.1(c)).

The above discussion can be extended to any given regression dataset where the aim is to construct an ϵ-insensitive regressor. We add and subtract ϵ' to y component of every training point, and obtain following two sets respectively

$$
\begin{aligned}
D^+ &= \left\{(x^{(i)}, y_i + \epsilon'), i = 1, \ldots, l\right\}, \\
D^- &= \left\{(x^{(i)}, y_i - \epsilon'), i = 1, \ldots, l\right\}.
\end{aligned} \tag{4.5}
$$

We next assign label $+1$ to every point of D^+ and -1 to every point of D^- and thus form a classification problem in R^{n+1} with $2l$ training points. Then the problem of constructing an ϵ'-insensitive hyperplane is equivalent to linearly separating these two sets. Infact, it has been shown that for given a regression training set, a regressor $y = x^T w + b$ is an ϵ-insensitive regressor if and only if the sets D^+ and D^- locate on the different sides of the (n+1) dimensional hyperplane $x^T w - y + b = 0$ respectively (Bi and Bennett [1]).

Let us apply SVM classification method to above two sets D^+ and D^- and assume that the classifying hyperplane is $x^T w + \eta y + b = 0$. We get the following QPP

$$
\begin{aligned}
&\underset{(w,\ b,\ \eta)}{\text{Min}} \quad \frac{1}{2}||w||^2 + \frac{1}{2}\eta^2 \\
&\text{subject to} \\
&\qquad Aw + \eta(Y + \epsilon' e) + be \geq 1, \\
&\qquad Aw + \eta(Y - \epsilon' e) + be \leq -1.
\end{aligned} \tag{4.6}
$$

Let $(w',\ b',\ \eta')$ be the solution to above formulation so that the classifier equation is $x^T w' + \eta' y + b' = 0$. Let η' be fixed and the transformations $w = -w/\eta'$, $b = -b/\eta'$, $\epsilon = \epsilon' - 1/\eta'$ be applied. Here it is to be noted that $\eta' \neq 0$ and $\epsilon > 0$. For these and other details we may refer to Bi and Bennett [1] and Deng et al. [17]. Get the resulting formulation as

$$
\begin{aligned}
&\underset{(w,\ b)}{\text{Min}} \quad \frac{1}{2}||w||^2 \\
&\text{subject to} \\
&\qquad Y - (Aw + eb) \leq \epsilon e, \\
&\qquad (Aw + eb) - Y \leq \epsilon e.
\end{aligned} \tag{4.7}
$$

The above formulation is same as that of ϵ-Support Vector Regression(SVR) applied to the original dataset. In this, we solve for $w \in R^n$ and $b \in R$ such that the regressor function given by $f(x) = x^T w + b$ is an ϵ-insensitive hyperplane.

It follows from the above transformations that $(-w'/\eta', -b'/\eta')$ is the solution to SVR formulation. Hence, if $x^T w + \eta y + b = 0$ is the classifier equation, then the regressor function is given by $f(x) = -x^T(w/\eta) - (b/\eta)$ which shows that the solution of the SVR problem is related to the SVM solution of resulting classification problem.

4.4 TWSVR via TWSVM

We now proceed to discuss the formulations of Goyal [16] and Khemchandani et al. [14, 15] twin support vector regression. Here the intuition is to derive TWSVR formulation by relating it to a suitable TWSVM formulation exactly in the same way as SVR formulation is related to SVM formulation. As a first step we construct D^+ and D^- as in Sect. 4.3 and get two hyperplanes for the resulting TWSVM problem. If we show that the mean of these hyperplanes linearly separates the two sets D^+ and D^-, then the mean becomes ϵ-insensitive hyperplane to the regression problem. This is true as TWSVM when applied on two sets A and B gives two hyperplanes $x^T w_1 + b_1 = 0$ and $x^T w_2 + b_2 = 0$ such that the shifted hyperplane $x^T w_1 + b_1 + 1 = 0$ lies above the points of set B and similarly the shifted hyperplane $x^T w_2 + b_2 - 1 = 0$ lies below the points of set A. Clearly then the mean of these shifted hyperplanes (which is same as mean of the original hyperplanes) will separate the two sets A and B.

Now the TWSVM methodology applied to two datasets D^+ and D^- produces following two QPPs yielding hyperplanes $x^T w_1 + \eta_1 y + b_1 = 0$ and $x^T w_2 + \eta_2 y + b_2 = 0$

$$\begin{aligned} \underset{(w_1,\, b_1,\, \eta_1,\, \xi_1)}{\text{Min}} \quad & \frac{1}{2}\|(Aw_1 + \eta_1(Y + \epsilon e) + eb_1)\|_2 + C_1 e^T \xi_1 \\ \text{subject to} \quad & \\ & -(Aw_1 + \eta_1(Y - \epsilon e) + eb_1) + \xi_1 \geq e, \\ & \xi_1 \geq 0, \end{aligned} \tag{4.8}$$

$$\begin{aligned} \underset{(w_2,\, b_2,\, \eta_2,\, \xi_2)}{\text{Min}} \quad & \frac{1}{2}\|(Aw_2 + \eta_2(Y - \epsilon e) + eb_2)\|_2 + C_2 e^T \xi_2 \\ \text{subject to} \quad & \\ & Aw_2 + \eta_2(Y + \epsilon e) + eb_2 + \xi_2 \geq e, \\ & \xi_2 \geq 0. \end{aligned} \tag{4.9}$$

Let us consider the first problem (4.8). Here we note that $\eta_1 \neq 0$ and therefore without any loss of generality, we can assume that $\eta_1 < 0$. We next consider the constraints of (4.8) and rewrite the same as

$$-\left[\left(A\left(\frac{w_1}{-\eta_1}\right) - (Y - \epsilon e) + e\left(\frac{b_1}{-\eta_1}\right)\right)(-\eta_1)\right] + \xi_1 \geq e, \xi_1 \geq 0. \tag{4.10}$$

On replacing $w_1 = -w_1/\eta_1$ and $b_1 = -b_1/\eta_1$ and noting that $-\eta_1 \geq 0$, we get

$$-(Aw_1 + eb_1) + \left(Y - e\left(\epsilon - \frac{1}{\eta_1}\right)\right) \geq -\left(\frac{\xi_1}{-\eta_1}\right), \xi_1 \geq 0. \tag{4.11}$$

Let $\epsilon_1 = \left(\epsilon - \frac{1}{\eta_1}\right)$. As $\eta_1 < 0$ and $\epsilon_1 > 0$, therefore from above

$$-(Y - \epsilon_1 e) + (Aw_1 + eb_1) \leq \xi_1, \xi_1 \geq 0, \tag{4.12}$$

where $\xi_1 = \left(\frac{\xi_1}{-\eta_1}\right) \geq 0$. We next consider the objective function of problem (4.8) and note that it can be re-written as

$$\frac{1}{2}\left\|\left(A\left(\frac{w_1}{-\eta_1}\right) - (Y + \epsilon e) + e\left(\frac{b_1}{-\eta_1}\right)\right)\right\|_2 \eta_1^2 + \frac{C_1}{-\eta_1}e^T\frac{\xi_1}{-\eta_1}\eta_1^2. \tag{4.13}$$

Again by replacing $w_1 = -w_1/\eta_1$, $b_1 = -b_1/\eta_1$, $\xi_1 := \left(\frac{\xi_1}{-\eta_1}\right) \geq 0$, and $C_1 = \left(\frac{C_1}{-\eta_1}\right) > 0$ we get the above objective function as

$$\eta_1^2\left[\frac{1}{2}\|(Aw_1 + eb_1 - (Y + \epsilon e))\|_2 + C_1 e^T \xi_1\right]. \tag{4.14}$$

In view of the above discussion, the problem (4.8) is equivalent to

$$\begin{aligned}
&\underset{(w_1,\, b_1,\, \xi_1)}{\text{Min}} \quad \frac{1}{2}\|(Y + e\epsilon) - (Aw_1 + eb_1)\|_2 + C_1 e^T \xi_1 \\
&\text{subject to} \\
&\qquad (Aw_1 + eb_1) - (Y - e\epsilon_1) \leq \xi_1, \\
&\qquad \xi_1 \geq 0.
\end{aligned} \tag{4.15}$$

Now we consider the objective function of problem (4.15) and write it as

$$\left\|Aw_1 + e\left(b_1 - \frac{1}{\eta_2}\right) - \left(\gamma + \left(\epsilon - \frac{1}{\eta_2}\right)e\right)\right\|_2 = \|Aw_1 + b_1 - (\gamma + e\epsilon_2)\|_2,$$

where $\epsilon_2 = (\epsilon - \frac{1}{\eta_2}) > 0$ and new $b_1 = e(b_1 - \frac{1}{\eta_2}) > 0$ is still denoted by b_1. This change requires a similar adjustment in the constraints. Thus

$$(Aw_1 + eb_1) - (\gamma - e\epsilon_1) \leq \xi_1,$$

is equivalent to

$$\left(Aw_1 + e\left(b_1 - \frac{1}{\eta_2}\right)\right) - (\gamma - e\epsilon_1) \leq \xi_1 - \frac{e}{\eta_2}.$$

but as the new $b_1 = e(b_1 - \frac{1}{\eta_2})$ is still denoted by b_1 and $\xi_1 = \xi_1 - \frac{\epsilon}{\eta_2} \geq 0$, $(\xi_1 \geq 0, \eta_2 < 0, \epsilon > 0)$ is still denoted by ξ_1, problem (4.15) is equivalent to (TWSVR1) stated below.

$$\text{(TWSVR1)} \qquad \underset{(w_1,\, b_1,\, \xi_1)}{\text{Min}} \quad \frac{1}{2}\|(Y + e\epsilon_2 - (Aw_1 + eb_1))\|_2 + C_1 e^T \xi_1$$

subject to

$$\begin{aligned} (Aw_1 + eb_1) - (Y - e\epsilon_1) &\leq \xi_1, \\ \xi_1 &\geq 0. \end{aligned} \tag{4.16}$$

In a similar manner, problem (4.9) is equivalent to

$$\text{(TWSVR2)} \qquad \underset{(w_2,\, b_2,\, \xi_2)}{\text{Min}} \quad \frac{1}{2}\|(Y - e\epsilon_1 - (Aw_2 + eb_2))\|_2 + C_2 e^T \xi_2$$

subject to

$$\begin{aligned} Y + e\epsilon_2 - (Aw_2 + eb_2) &\leq \xi_2, \\ \xi_2 &\geq 0. \end{aligned} \tag{4.17}$$

where $C_1, C_2 \geq 0$, ξ_1 and ξ_2 are error vectors and e is a vector of ones of appropriate dimension. Each of the above two QPP gives function $f_1(x) = x^T w_1 + b_1$ and $f_2(x) = x^T w_2 + b_2$ respectively. The final regressor function is obtained by taking the mean of the above two functions.

The two formulations (TWSVR1) and (TWSVR2) yield two hyperplanes which geometrically bounds the dataset from up and down. We can understand the bounding regressors as forming the epsilon tube within which the data resides. Specifically, the function $f_1(x)$ determines the ϵ_1 insensitive down bound regressor and the function $f_2(x)$ determines the ϵ_2 insensitive up bound regressor. The final regressor is obtained by taking mean of up and down bound regressor functions. Figure 4.2 shows the geometrical interpretation of the two bounding regressors. The first term in the objective function of (TWSVR1) and (TWSVR2) is the sum of squared distance from the shifted function $y = x^T w_1 + b_1 - \epsilon_2$ and $y = x^T w_2 + b_2 + \epsilon_1$ respectively to the training points. Minimizing it will keep the regressor function $f_1(x)$ and $f_2(x)$ close to the training dataset. The constraints of (TWSVR1) and (TWSVR2) requires $f_1(x)$ and $f_2(x)$ to be at least ϵ_1 down and ϵ_2 up respectively from the training points. The second term in the objective function minimizes the sum of error variables, trying to overfit the training dataset by not allowing training points to come ϵ_1 close to down bound regressor or ϵ_2 close to up bound regressor.

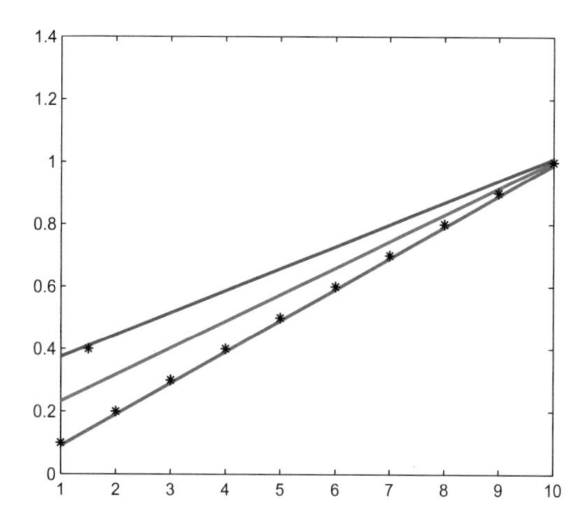

Fig. 4.2 Geometrical interpretation of bounding regressors and epsilon tube

Solving two formulations instead of one as in classical SVR provides the advantage that the bounding regressors which are now obtained by solving two different formulations and not one are no longer required to be parallel as in case of SVR. Since these bounding regressors which forms the epsilon tube are no longer required to be parallel can become zero at some places at non zero at other. Figure 4.2 again shows this interpretation of epsilon tube. This makes the final regressor to generalize well as compared to SVR. Similar advantage is observed in TWSVM as compared to SVM. Also the two formulations of TWSVR are each smaller than the formulation of SVR because the constraints of the SVR formulation splits to two formulations of TWSVR making the number of inequalities in the constraints of TWSVR smaller and hence making the TWSVR model faster. Similar to TWSVM formulation, we obtain that TWSVR is approximately four times faster than standard SVR. These advantages of TWSVR over SVR in terms of performance as well as computational time makes them better suited for a regression problem.

Next on the lines of Shao et al. [18], we introduce a regularization term in each of the two TWSVR formulation to control for the Structural Risk and from now on we call the following two formulations as TWSVR formulations

$$
\begin{aligned}
(\text{TWSVR1}) \quad \underset{(w_1,\, b_1,\, \xi_1)}{\text{Min}} \quad & C_1 e^T \xi_1 + \frac{C_3}{2}(\|w_1\|_2 + (b_1)^2) + \frac{1}{2}\|(Y + e\epsilon_2 - (Aw_1 + eb_1))\|_2 \\
\text{subject to} \quad & \\
& (Aw_1 + eb_1) - (Y - e\epsilon_1) \leq \xi_1, \\
& \xi_1 \geq 0. \qquad (4.18)
\end{aligned}
$$

In a similar manner, problem (4.8) is equivalent to

$$\text{(TWSVR2)} \quad \underset{(w_2,\, b_2,\, \xi_2)}{\text{Min}} \quad C_2 e^T \xi_2 + \frac{C_4}{2}(\|w_2\|_2 + (b_2)^2) + \frac{1}{2}\|(Y - e\epsilon_1 - (Aw_2 + eb_2))\|_2$$

subject to

$$\begin{aligned} Y + e\epsilon_2 - (Aw_2 + eb_2) &\leq \xi_2, \\ \xi_2 &\geq 0, \end{aligned} \tag{4.19}$$

where $C_3, C_4 > 0$ are regularization parameters.

4.4.1 A Dual Formulation of TWSVR

As with (TWSVM), we work with the dual formulation of TWSVR. For this we consider the Lagrangian corresponding to the problem(TWSVR1) and write its Wolfe's dual. The Lagrangian function is

$$\begin{aligned} L(w_1, b_1, \xi_1, \alpha, \beta) = {} & C_1 e^T \xi_1 + \alpha^T ((Aw_1 + eb_1) - (Y - e\epsilon_1) - \xi_1)) \\ & + \frac{1}{2}((Y + e\epsilon_2 - (Aw_1 + eb_1))^T \times (Y + e\epsilon_2 - (Aw_1 + eb_1))) \\ & \frac{C_3}{2}(w_1^T w_1 + b_1^2) - \beta^T \xi_1, \end{aligned} \tag{4.20}$$

where $\alpha = (\alpha_1, \alpha_2 \dots \alpha_l)^T$, and $\beta = (\beta_1, \beta_2 \dots \beta_l)^T$ are the vectors of Lagrange multipliers. The Karush–Kuhn–Tucker (K. K. T) necessary and sufficient optimality conditions (Mangasarian [19] and Chandra et al. [20]) for (TWSVR1) are given by

$$C_3 w_1 - A^T(Y + e\epsilon_2 - (Aw_1 + eb_1)) + A^T\alpha = 0, \tag{4.21}$$

$$C_3 b_1 - e^T(Y + e\epsilon_2 - (Aw_1 + eb_1)) + e^T\alpha = 0, \tag{4.22}$$

$$C_1 e - \alpha - \beta = 0, \tag{4.23}$$

$$(Aw_1 + eb_1 - (Y - e\epsilon_1)) \leq \xi_1, \quad \xi_1 \geq 0, \tag{4.24}$$

$$\alpha^T((Aw_1 + eb_1) - (Y - e\epsilon_1) - \xi_1)) = 0, \quad \beta^T \xi_1 = 0, \tag{4.25}$$

$$\alpha \geq 0, \quad \beta \geq 0. \tag{4.26}$$

Since $\beta \geq 0$, from (4.23) we have

$$0 \leq \alpha \leq C_1. \tag{4.27}$$

Next, combining (4.21) and (4.22) leads to

$$C_3\begin{bmatrix} w_1 \\ b_1 \end{bmatrix} + [-A^T \quad -e^T]\left(Y + e\epsilon_2 - [A \quad e]\begin{bmatrix} w_1 \\ b_1 \end{bmatrix}\right) + [A^T \quad e^T]\alpha = 0. \quad (4.28)$$

We define

$$H = [A \quad e], \quad (4.29)$$

and the augmented vector $u = \begin{bmatrix} w_1 \\ b_1 \end{bmatrix}$. With these notations, (4.28) may be rewritten as

$$C_3 u - H^T(Y + e\epsilon_2 - Hu) + H^T\alpha = 0, \quad (4.30)$$

giving

$$-H^T(Y + e\epsilon_2) + (C_3 I + H^T H)u + H^T\alpha = 0, \quad (4.31)$$

where I is an identity matrix of appropriate dimension. We further get

$$u = (H^T H + C_3 I)^{-1}(H^T(Y + e\epsilon_2) - H^T\alpha). \quad (4.32)$$

Here we note that if $C_3 = 0$, i.e., there is no regularization term in the objective function of TWSVR formulations, then it is possible that $H^T H$ may not be well conditioned and hence on lines of (Saunders et al. [21]), we would be required to introduce explicitly a regularization term $\delta I,\ \delta > 0$, to take care of problems due to possible ill-conditioning of $H^T H$. However because of the addition of the extra regularization term in the objective function of TWSVR on the lines of TWSVM (see Chap. 3), this problem of ill-conditioning is overcome and no measures need to taken explicitly.

Using (4.20) and the above K.K.T conditions, we obtain the Wolfe dual (Mangasarian [19], Chandra et al. [20]) (DTWSVR1) of (TWSVR1) as follows

$$\text{(DTWSVR1)} \qquad \underset{\alpha}{\text{Min}} \ -\frac{1}{2}\alpha^T H(H^T H + C_3 I)^{-1}H^T\alpha + f^T H(H^T H + C_3 I)^{-1}H^T\alpha$$
$$- f^T\alpha + (\epsilon_1 + \epsilon_2)e^T\alpha$$

subject to

$$0 \le \alpha \le C_1 e. \quad (4.33)$$

Similarly, we consider (TWSVR2) and obtain its dual as (DTWSVM2)

$$\text{(DTWSVR2)} \qquad \underset{\gamma}{\text{Min}} \ g^T\gamma + (\epsilon_1 + \epsilon_2)e^T\gamma - \frac{1}{2}\gamma^T H(H^T H + C_4 I)^{-1}H^T\gamma$$
$$- g^T H(H^T H + C_4 I)^{-1}H^T\gamma$$

subject to

$$0 \le \gamma \le C_2 e, \quad (4.34)$$

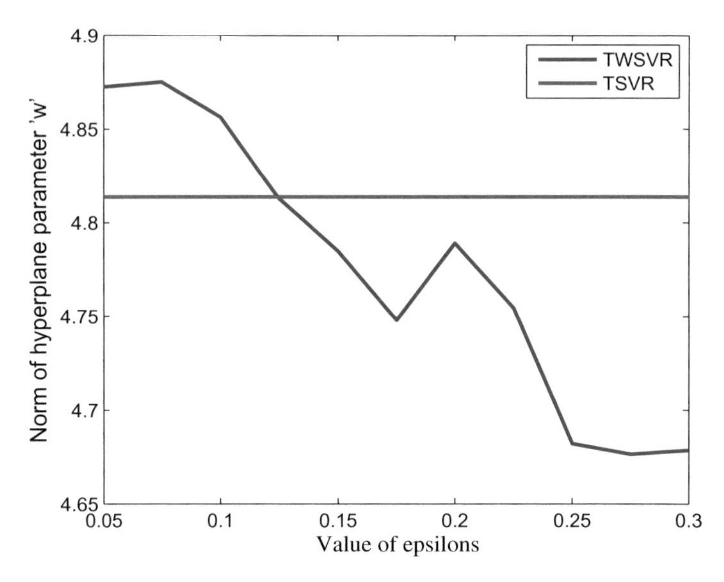

Fig. 4.3 TWSVR and TSVR plot of norm of w hyperplane parameter versus different values of $\epsilon's$ for Servo dataset and $\epsilon_1 = \epsilon_2$

where $H = [A, \quad e], \quad f = Y + e\epsilon_2; \quad g = Y - e\epsilon_1, \quad u_1 = [w_1, b_1]^T = (H^T H + C_3 I)^{-1} H^T (f - \alpha), u_2 = [w_2; b_2] = (H^T H + C_4 I)^{-1} H^T (g + \gamma)$. Once the vectors u_1 and u_2 are known, the two functions $f_1(x)$ and $f_2(x)$ are obtained.

We shall now like to make following observations on our TWSVR model (TWSVR1)–(TWSVR2).

(i) TWSVR formulation is really in the true spirit of TWSVM because the objective function of (TWSVR1) consists of data points having y component as $y + \epsilon e$, and the constraints consists of those data points for which y component is $y - \epsilon e$. This is really the essence of twin spirit which is missing in Peng's formulation.
(ii) Our formulation, unlike Peng's formulation, inherently depends on values of ϵ_1 and ϵ_2. These values are responsible for the orientation of the final regressor.
(iii) Unlike Peng's formulation which is mostly adhoc, our formulations are mathematically derived. Because our TWSVR formulation carries the essence of twin spirit it should be a better choice for future research in this area.
(iv) As mentioned earlier, we have performed an experiment on Servo dataset (Blake and Merz [22]) and plotted the norm of hyperplane parameter w versus different values of $\epsilon (= \epsilon_1 = \epsilon_2)$ in Fig. 4.3. We can observe that the w parameter is independent of values of $\epsilon's$ for the Peng's TSVR model, unlike in TWSVR model. Further, we have also checked the variation of Sum Square Error (SSE) with different values of $\epsilon (= \epsilon_1 = \epsilon_2)$ for the sinc function. The plots obtained via TSVR, TWSVR and original data has been depicted in Fig. 4.4 and corresponding relation between the SSE and ϵ has been shown in Fig. 4.5.

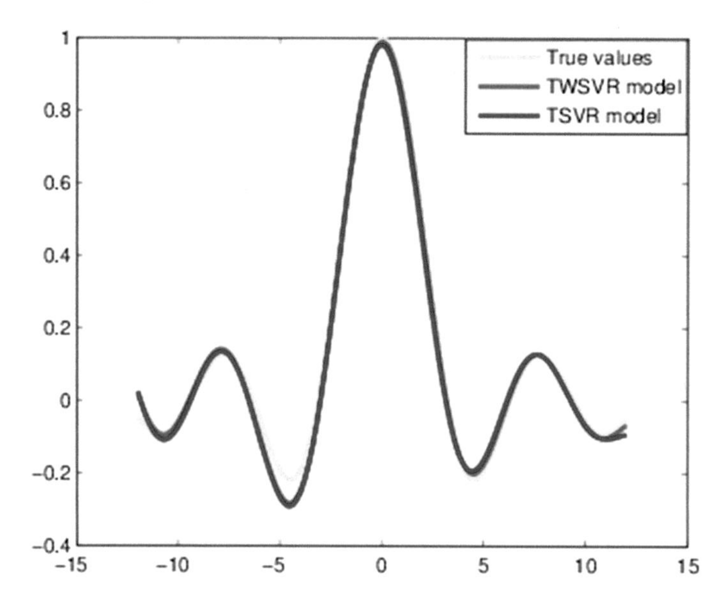

Fig. 4.4 TWSVR and TSVR plot of Sinc function

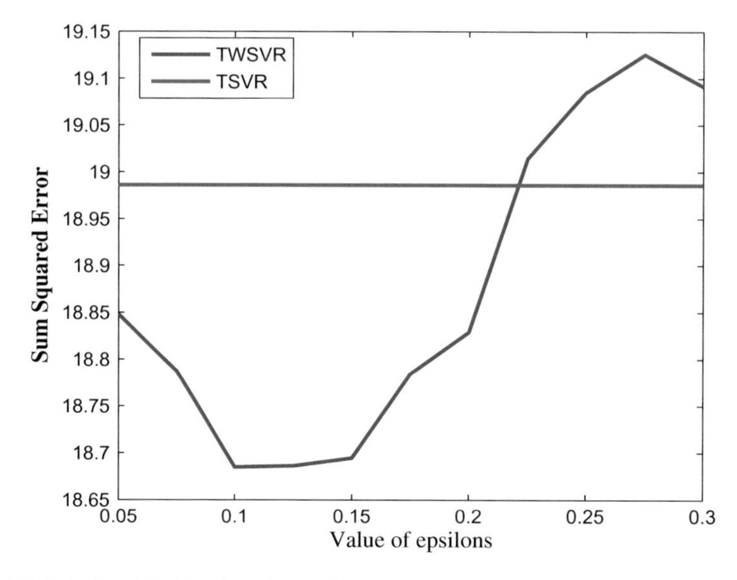

Fig. 4.5 TWSVR and TSVR plot of sum of square error versus different values of $\epsilon's$ for Sinc function

4.4.2 Kernel Version of TWSVR

We now state the primal and dual versions of Kernel TWSVR which can be derived on the lines of Kernel TWSVM discussed in Chap. 3. We consider the following kernel generated functions instead of linear functions.

$$f_1(x) = K(x^T, A^T)w_1 + b_1, \quad \text{and} \quad f_2(x) = K(x^T, A^T)w_2 + b_2, \tag{4.35}$$

where K is the kernel matrix corresponding to an appropriately chosen kernel function.

The motivation for choosing the kernel generated surfaces in the form of (4.35) is that by choosing $K(x^T, A^T) = x^T A$ and defining $w_1 := A^T w_1$, the hyperplanes corresponding to linear TWSVR are obtained as a special case of (4.35). Therefore, in line with the arguments in Sect. 4.4, we construct optimization problems (KTWSVR1) and (KTWSVR2) as follows

$$\text{(KTWSVR1)} \quad \underset{(w_1,\, b_1,\, \xi_1)}{\text{Min}} \quad C_1 e^T \xi_1 + \frac{1}{2}\|(Y + e\epsilon_2 - (K(A, A^T)w_1 + eb_1))\|_2$$
$$\frac{C_3}{2}(\|w_1\|_2 + (b_1)^2)$$
$$\text{subject to}$$
$$(K(A, A^T)w_1 + eb_1) - (Y - e\epsilon_1) \le \xi_1,$$
$$\xi_1 \ge 0. \tag{4.36}$$

In a similar manner, problem (4.8) is equivalent to

$$\text{(KTWSVR2)} \quad \underset{(w_2,\, b_2,\, \xi_2)}{\text{Min}} \quad C_2 e^T \xi_2 + \frac{1}{2}\|(Y - e\epsilon_1 - (K(A, A^T)w_2 + eb_2))\|_2$$
$$+\frac{C_4}{2}(\|w_2\|_2 + (b_2)^2)$$
$$\text{subject to}$$
$$Y + e\epsilon_2 - (K(A, A^T)w_2 + eb_2) \le \xi_2,$$
$$\xi_2 \ge 0. \tag{4.37}$$

We next consider the Lagrangian corresponding to the problem(KTWSVR1) and write its Wolfe dual. The Lagrangian function is

$$L(w_1, b_1, \xi_1, \alpha, \beta) = \alpha^T((K(A, A^T)w_1 + eb_1) - (Y - e\epsilon_1) - \xi_1)) - \beta^T \xi_1 + C_1 e^T \xi_1$$
$$+\frac{1}{2}((Y + e\epsilon_2 - (K(A, A^T)w_1 + eb_1))^T \times (Y + e\epsilon_2 - (K(A, A^T)w_1 + eb_1)))$$
$$+\frac{C_3}{2}((w_1)^T(w_1) + (b_1)^2), \tag{4.38}$$

where $\alpha = (\alpha_1, \alpha_2 \ldots \alpha_l)^T$, and $\beta = (\beta_1, \beta_2 \ldots \beta_l)^T$ are the vectors of Lagrange multipliers. The Karush–Kuhn–Tucker (K. K. T) necessary and sufficient optimality conditions (Mangasarian [19]) for (KTWSVR1) are given by

$$C_3 w_1 - K(A, A^T)^T (Y + e\epsilon_2 - (K(A, A^T)w_1 + eb_1)) + K(A, A^T)^T \alpha = 0, \tag{4.39}$$

$$C_3 b_1 - e^T (Y + e\epsilon_2 - (K(A, A^T)w_1 + eb_1)) + e^T \alpha = 0, \tag{4.40}$$

$$C_1 e - \alpha - \beta = 0, \tag{4.41}$$

$$(K(A, A^T)w_1 + eb_1 - (Y - e\epsilon_1)) \leq \xi_1, \quad \xi_1 \geq 0, \tag{4.42}$$

$$\alpha^T((K(A, A^T)w_1 + eb_1) - (Y - e\epsilon_1) - \xi_1)) = 0, \quad \beta^T \xi_1 = 0, \tag{4.43}$$

$$\alpha \geq 0, \quad \beta \geq 0. \tag{4.44}$$

Since $\beta \geq 0$, from (4.41) we have

$$0 \leq \alpha \leq C_1. \tag{4.45}$$

Next, combining (4.39) and (4.40) leads to

$$C_3 \begin{bmatrix} w_1 \\ b_1 \end{bmatrix} + \begin{bmatrix} -K(A, A^T)^T & -e^T \end{bmatrix} (Y + e\epsilon_2 - [K(A, A^T) \;\; e]) \begin{bmatrix} w_1 \\ b_1 \end{bmatrix} + [K(A, A^T)^T \;\; e^T]\alpha = 0. \tag{4.46}$$

We define

$$H = [K(A, A^T) \;\; e], \tag{4.47}$$

and the augmented vector $u = \begin{bmatrix} w_1 \\ b_1 \end{bmatrix}$. With these notations, (4.46) may be rewritten as

$$C_3 u - H^T (Y + e\epsilon_2 - Hu) + H^T \alpha = 0, \tag{4.48}$$

giving

$$-H^T (Y + e\epsilon_2) + (C_3 I + H^T H)u + H^T \alpha = 0, \tag{4.49}$$

where I is an identity matrix of appropriate dimension. We further get

$$u = (H^T H + C_3 I)^{-1}(H^T(Y + e\epsilon_2) - H^T \alpha). \tag{4.50}$$

Using (4.38) and the above K.K.T conditions, we obtain the Wolfe dual (Mangasarian [19] and Chandra et al. [20]) (DKTWSVR1) of (DKTWSVR1) as follows

$$\text{(DKTWSVR1)} \quad \underset{\alpha}{\text{Min}} \; -f^T\alpha + (\epsilon_1 + \epsilon_2)e^T\alpha - \frac{1}{2}\alpha^T H(H^T H + C_3 I)^{-1} H^T \alpha$$
$$+ f^T H(H^T H + C_3 I)^{-1} H^T \alpha$$

subject to

$$0 \leq \alpha \leq C_1 e. \tag{4.51}$$

Similarly, we consider (KTWSVR2) and obtain its dual as (DKTWSVR2)

$$\text{(DKTWSVR2)} \quad \underset{\gamma}{\text{Min}} \; g^T\gamma + (\epsilon_1 + \epsilon_2)e^T\gamma - \frac{1}{2}\gamma^T H(H^T H + C_4 I)^{-1} H^T \gamma$$
$$- g^T H(H^T H + C_4 I)^{-1} H^T \gamma$$

subject to

$$0 \leq \gamma \leq C_2 e, \tag{4.52}$$

where $H = [K(A, A^T), \;\; e]$; $f = Y + e\epsilon_2$; $g = Y - e\epsilon_1$; $u_1 = [w_1, b_1]^T = (H^T H + C_3 I)^{-1} H^T(f - \alpha)$, $u_2 = [w_2; b_2] = (H^T H + C_4 I)^{-1} H^T(g + \gamma)$. Once the vectors u_1 and u_2 are known, the two functions $f_1(x)$ and $f_2(x)$ are obtained.

4.4.3 Experiments

The SVR, TWSVR and TSVR models were implemented by using MATLAB 7.8 running on a PC with Intel Core 2 Duo processor (2.00 GHz) with 3 GB of RAM. The methods were evaluated on two artificial dataset and on some standard regression datasets from UCI Machine Learning Laboratory (Blake and Merz [22]).

For our simulations, we have considered RBF Kernel and the values of parameters like C(regularization term), epsilon and sigma are selected from the set of values $\{2^i | i = -9, -8, \ldots, 10\}$ by tuning a set comprising of random 10 percent of sample set. Similar to Peng [3], we have taken $C_1 = C_2$ and $\epsilon = \epsilon_1 = \epsilon_2$ in our experiments to degrade the computational complexity of parameter selection.

The datasets used for comparison and the evaluation criteria is similar to Peng [3], further for the sake of completeness we would specify the evaluation criteria before presenting the experimental results. The total number of testing samples is denoted by l, y_i denotes the real value of a sample x_i, $\widehat{y_i}$ denotes the predicted value of sample

x_i and $\overline{y} = \Sigma_{i=1}^{l} y_i$ is the mean of $y_1, y_2, \ldots, y_l$. We use the following criterion for algorithm evaluation.

SSE: Sum squared error of testing, defines as SSE $= \sum_{i=1}^{l} (y_i - \widehat{y_i})^2$.

SST: Sum squared deviation of testing, which is defined as SST $= \sum_{i=1}^{l} (y_i - \overline{y})^2$

SSR: Sum squared deviation that can be explained by the estimator, which is defines as SSR $= \sum_{i=1}^{l} (\widehat{y_i} - \overline{y})^2$. It reflects the explanation ability of the regressor.

SSE/SST: Ratio between sum squared error and sum squared deviation of testing samples, which is defines as SSE/SST $= \sum_{i=1}^{l} (y_i - \widehat{y_i})^2 / \sum_{i=1}^{l} (y_i - \overline{y})^2$.

SSR/SST: Ratio between interpretable sum squared deviation and real sum squared deviation of testing samples, which is defined as SSR/SST $= \sum_{i=1}^{l} (\widehat{y_i} - \overline{y})^2 / \sum_{i=1}^{l} (y_i - \overline{y})^2$

In most cases, small SSE/SST means good agreement between estimations and real values, and to obtain smaller SSE/SST usually accompanies an increase of SSR/SST. However, extremely small value of SSE/SST indicate overfitting of the regressor. Therefore a good estimator should strike a balance between SSE/SST and SSR/SST. The other two criterion's are SSE/SST^{LOO} and SSR/SST^{LOO}, which are defined as SSE/SST^{L00} $= \sum_{i=1}^{l} (y_i - \hat{y}_i^{LOO})^2 / \sum_{i=1}^{l} (y_i - \overline{y}^{tr})^2$ and SSR/SST^{LOO} $= \sum_{i=1}^{l} (\hat{y}_i^{LOO} - \overline{y}^{tr})^2 / \sum_{i=1}^{l} (y_i - \overline{y}^{tr})^2$, where $\hat{y}_i^{LOO}$ is the prediction of y_i when a sample x_i is left out from training set during leave one out procedure and $\overline{y}^{tr}$ is the mean of y values of m training data points.

We first compare the performance of our proposed TWSVR on synthetic dataset Sinc function, which is defined as

$$y = \text{Sinc}(x) = \frac{\sin(x)}{x}, \quad \text{where} \quad x \in [-12, 12]. \tag{4.53}$$

To effectively reflect the performance of our method, training data points are perturbed by different Gaussian noises with zero means. Specifically, we have the following training samples (x_i, y_i), $i = 1, \ldots, l$.

(Type A)

$$y_i = \frac{sin(x_i)}{x_i} + \xi_i, \tag{4.54}$$

where $x_i \sim U[-12, 12]$ and $\xi_i \sim N(0, 0.1^2)$.
(Type B)

$$y_i = \frac{sin(x_i)}{x_i} + \xi_i, \tag{4.55}$$

where $x_i \sim U[-12, 12]$ and $\xi_i \sim N(0, 0.2^2)$. Here $U[a, b]$ and $N(c, d)$ represent the Uniform random variable in $[a, b]$ and the Gaussian random variable with mean c and variance d^2, respectively.

Next, we compare TWSVR on the following synthetic dataset

$$g(x) = \frac{|x-1|}{4} + |sin(\pi(1 + (x-1)/4))| + 1, \tag{4.56}$$

where $x \in [-10, 10]$. Again the training samples are polluted with Gaussian noises with zero means and different variances as follows
(Type C)

$$y_i = g(x_i) + \xi_i, \tag{4.57}$$

where $x_i \sim U[-10, 10]$ and $\xi_i \sim N(0, 0.2^2)$.
(Type D)

$$y_i = g(x_i) + \xi_i, \tag{4.58}$$

where $x_i \sim U[-10, 10]$ and $\xi_i \sim N(0, 0.4^2)$. To avoid biased comparisons, we generate 10 groups of noisy samples, which respectively consists of 256(400) training samples for Type A and B (Type C and D) and 500(800) test samples for Type A and B (Type C and D). Besides, testing data points are uniformly sampled without considering any noise. Tables 4.1 and 4.2 shows the average results of TWSVR, TSVR and SVR with 10 independent runs on aforementioned synthetic datasets.

Table 4.1 Result comparison of TWSVR, Peng's TSVR and classical SVR on Sinc dataset with different noises

Noise	Regressor	SSE	SSE/SST	SSR/SST
Type A	**TWSVR**	0.1926 ± 0.0796	0.0034 ± 0.0014	0.9959 ± 0.0281
	TSVR	0.2511 ± 0.0851	0.0045 ± 0.0015	1.0020 ± 0.0328
	SVR	0.2896 ± 0.1355	0.0052 ± 0.0024	0.9865 ± 0.0444
Type B	**TWSVR**	0.7671 ± 0.3196	0.0137 ± 0.0057	0.9940 ± 0.0607
	TSVR	0.9527 ± 0.3762	0.0170 ± 0.0067	1.0094 ± 0.0579
	SVR	1.0978 ± 0.3570	0.0196 ± 0.0064	1.0187 ± 0.0818

Table 4.2 Result comparison of TWSVR, Peng's TSVR and classical SVR on g(x) function with different noises

Noise	Regressor	SSE	SSE/SST	SSR/SST
Type C	**TWSVR**	2.5632 ± 0.4382	0.0048 ± 0.0008	0.9967 ± 0.0185
	TSVR	2.8390 ± 0.7234	0.0053 ± 0.0014	0.9992 ± 0.0175
	SVR	3.6372 ± 0.7828	0.0068 ± 0.0015	0.9832 ± 0.0293
Type D	**TWSVR**	8.5473 ± 1.9399	0.0160 ± 0.0036	1.0295 ± 0.0560
	TSVR	10.0011 ± 1.8801	0.0187 ± 0.0035	1.0230 ± 0.0572
	SVR	9.4700 ± 2.1139	0.0177 ± 0.0040	1.0195 ± 0.0521

Table 4.3 Result comparison of our model of TWSVR, Peng's model of TSVR and classical SVR on Motorcycle and Diabetes datasets

Dataset	Regression type	SSE/SST^{100}	SSR/SST^{100}
Motorcycle	TWSVR	0.2280	0.7942
	TSVR	0.2364	0.8048
	SVR	0.2229	0.8950
Diabetes	**TWSVR**	0.6561	0.5376
	TSVR	0.6561	0.5376
	SVR	0.6803	0.3231

Next, we have shown the comparison on UCI benchmark dataset, i.e. we have tested the efficacy of the algorithms on several benchmark datasets which includes Motorcycle, Diabetes, Boston Housing, Servo, Machine CPU, Auto Price and Wisconsin B.C. For these datasets we have reported ten fold cross validation result with RBF kernel, except Motorcycle and Diabetes where we have used leave one out cross validation approach and Wisconsin B.C where we have used the linear kernel.

Table 4.3 shows the comparison results for Motorcycle and Diabetes datasets while Table 4.4 shows the comparison results for other five UCI datasets. We make the following observations from the results of comparison tables

(i) The TWSVR model derives the smallest SSE and SSE/SST for almost all the datasets when compared to SVR. This clearly shows the superiority of TWSVR over SVR in terms of testing accuracy.
(ii) The TWSVR performs better than TSVR when compared over synthetic datasets while they perform similar in case of UCI datasets. We note here that although both the models TWSVR and TSVR differ significantly conceptually, there dual optimization problems differ only in terms of epsilon values. In practice, ϵ is set to small values and hence both the models achieves similar results.

Shao et al. [5] presented another formulation termed as $\epsilon - TSVR$. The formulation of $\epsilon - TSVR$ is different from that of Peng [3] and is motivated by the classical SVR formulation. Specifically, Shao et al. [5] constructed the following pair of quadratic

Table 4.4 Result comparison of our model of TWSVR, Peng's model of TSVR and classical SVR on following UCI datasets

Dataset	Regression type	SSE/SST	SSR/SST
Machine CPU	**TWSVR**	0.0048 ± 0.0040	0.9805 ± 0.0984
	TSVR	0.0050 ± 0.0044	0.9809 ± 0.1004
	SVR	0.0325 ± 0.0275	0.7973 ± 0.1804
Servo	**TWSVR**	0.1667 ± 0.0749	0.9639 ± 0.2780
	TSVR	0.1968 ± 0.1896	1.0593 ± 0.4946
	SVR	0.1867 ± 0.1252	0.7828 ± 0.2196
Boston Housing	**TWSVR**	0.1464 ± 0.079	0.9416 ± 0.1294
	TSVR	0.1469 ± 0.079	0.9427 ± 0.13
	SVR	0.1546 ± 0.053	0.8130 ± 0.1229
Auto Price	**TWSVR**	0.1663 ± 0.0806	0.9120 ± 0.372
	TSVR	0.1663 ± 0.0806	0.9120 ± 0.372
	SVR	0.5134 ± 0.1427	0.3384 ± 0.1171
Wisconsin B.C	TWSVR	0.9995 ± 0.2718	0.4338 ± 0.2248
	TSVR	0.9989 ± 0.2708	0.4330 ± 0.2242
	SVR	0.9113 ± 0.1801	0.2398 ± 0.1294

programming problems to determine the down bound and up bound regressor $f_1(x) = w_1^T x + b_1$ and $f_2(x) = w_2^T x + b_2$ respectively,

$$(\epsilon - TSVR1) \qquad \underset{(w_1,\, b_1,\, \xi_1)}{\text{Min}} \quad C_1 e^T \xi_1 + \frac{1}{2}\|(Y - (Aw_1 + eb_1)\|_2 \frac{C_3}{2}(\|w_1\|_2 + (b_1)^2)$$

subject to

$$\begin{aligned} (Aw_1 + eb_1) - Y &\leq \xi_1 + e\epsilon_1, \\ \xi_1 &\geq 0, \end{aligned} \tag{4.59}$$

and

$$(\epsilon - TSVR2) \qquad \underset{(w_2,\, b_2,\, \xi_2)}{\text{Min}} \quad C_2 e^T \xi_2 + \frac{1}{2}\|(Y - (Aw_2 + eb_2)\|_2 + \frac{C_4}{2}(\|w_2\|_2 + (b_2)^2)$$

subject to

$$\begin{aligned} Y - (Aw_2 + eb_2) &\leq \xi_2 + e\epsilon_2, \\ \xi_2 &\geq 0. \end{aligned} \tag{4.60}$$

Here $C_1, C_2, C_3, C_4 > 0$ are regularization parameters and $\epsilon_1, \epsilon_2 > 0$.

Once $f_1(x)$ and $f_2(x)$ are determined the final estimated regression is again constrained as

$$f(x) = \frac{1}{2}(f_1(x) + f_2(x)) = \frac{1}{2}(w_1 + w_2)^T x + \frac{1}{2}(b_1 + b_2).$$

Shao et al. [5] also presented the kernel version of $\epsilon - TSVR$ (Kernel $\epsilon - TSVR$) and did extensive numerical experimentation to test the efficiency of their formulation.

One natural question here is to check if the $\epsilon - TSVR$ formulation of Shao et al. [5] can also be derived by Bi and Bennett [1] result. This is in fact true as has been demonstrated in Goyal [16].

4.5 Simultaneous Learning of Function and Its Derivative

A wide variety of problems with applications in diverse areas such as computer graphics, image processing etc. use estimation of a function along with its derivatives (Zheng [23]; Lagaris et al. [24]; Antonio et al. [25]; Julier and Uhlmann [26]). Also many times, it is easy to obtain the derivative information compared to functional value information. As an example, consider determining the velocity of a moving object compared to its absolute position. Additionally, a technique that uses information of function as well as derivatives can benefit by requiring fewer samples of function points to obtain a representation with a desired accuracy.

Recently, Lázarao et al. [27, 28] proposed a SVM based approach for the simultaneous learning of a function and its derivatives which uses an iterative weighted least squares (IRWLS) procedure. On similar lines, Jayadeva et al. [29, 30], also provided a solution to the above-mentioned problem of simultaneous learning of a function and its derivatives using a regularized least squares support vector machine framework (RLSVRD). The algorithm obtains the solution by inverting a single positive definite matrix instead of solving a quadratic programming problem (QPP).

Later in the spirit of Twin Support Vector Regression(TSVR) (Peng [3]), Khemchandani et al. [31] proposed nonparallel plane based support vector regression algorithm for the simultaneous learning of a function and its derivatives, which they termed as TSVRD. The algorithm solves two QPPs corresponding to the upper and lower bound ϵ-tubes around the function values and its derivatives. Experiment results on several functions of more than one variable (Khemchandani et al. [31]) show the superiority of the TSVRD against the above mentioned RLSVRD and Lazaro's approach.

Finally, Goyal [16] proposed a nonparallel plane based regressor for the simultaneous learning of a function and its derivatives. In the spirit of GEPSVR (Khemchandani et al. [32]), the algorithm requires solving of a pair of generalized eigenvalue problems to determine a pair of ϵ-insensitive bounding regressor around the function values and its derivatives. Experiment results over several functions of more than one variable indicate that the proposed method by Goyal [16] is superior to the aforementioned approaches in terms of testing accuracy. The proposed method is also fast when compared to TSVRD and Lazaro's approach as it requires the solution of generalized eigenvalue problem as compared to solution of QPP required by other two approaches.

4.5.1 Existing Variants

1. The SVR Approach of Lazaro et al.

Consider a set of M data samples $\{(x^{(i)}, y_i, y_i^{'}) \in \mathbb{R} \times \mathbb{R} \times \mathbb{R}, i = 1, 2, \ldots, l\}$, where $x^{(i)}$ denotes the ith sample, and y_i (respectively, $y_i^{'}$) denotes the value of the function (respectively, its derivative) at the ith sample. The proposed SVR model of Lázarao et al. [27, 28] for the simultaneous learning of a real valued function and its derivative determines a regressor of the form $f(x) = w^T\phi(x) + b$, where $\phi : \mathbb{R} \to \mathbb{R}^d$, and $d > 1$ is the dimension of a suitable feature space. Lázarao et al. [27, 28] proposed the following optimization problem for finding the desired regressor parameters w and b

$$\underset{(p_1,p_2,q_1,q_2,w,b)}{\text{Min}} \quad \frac{1}{2}w^T w + c_1(e^T p_1 + e^T p_2) + c_2(e^T q_1 + e^T q_2)$$

subject to

$$w^T\phi(x^{(i)}) + b - y_i \leq \epsilon + q_{1i}, \quad (i = 1, 2, \ldots, l), \tag{4.61}$$

$$y_i - w^T\phi(x^{(i)}) - b \leq \epsilon + q_{2i}, \quad (i = 1, 2, \ldots, l), \tag{4.62}$$

$$w^T\phi^{'}(x^{(i)}) - y_i^{'} \leq \epsilon_1 + p_{1i}, \quad (i = 1, 2, \ldots, l), \tag{4.63}$$

$$y_i^{'} - w^T\phi^{'}(x^{(i)}) \leq \epsilon_1 + p_{2i}, \quad (i = 1, 2, \ldots, l), \tag{4.64}$$

$$q_{1i}, q_{2i}, p_{1i}, p_{2i} \geq 0, \quad (i = 1, 2, \ldots, l), \tag{4.65}$$

where $\phi^{'}(x)$ denotes the derivative of $\phi(x)$; ϵ, $\epsilon_1 > 0$; c_1, c_2 are parameters and $p_{1i}, p_{2i}, q_{1i}, q_{2i}$, $(i = 1, 2, \ldots, l)$ are non-negative error variables. The solution of the problems (4.61)–(4.65) can be obtained by writing the Wolfe's dual and solving the same by using the Quadratic programming (QP) techniques but since the solution of the QPP becomes computationally expensive for large datasets, IRWLS procedure (Lázarao et al. [27, 28]) had also been developed. Finally we note that this method requires functional and derivative values at all l points.

2. RLSVRD: Regularized Least Squares Support Vector Regression of a Function and Its Derivatives

For a real valued function $f : \mathbb{R}^n \to \mathbb{R}$, consider the dataset of functional values $(x^{(i)}, y_i) \in \mathbb{R}^n \times \mathbb{R}$, $(i = 1, 2, \ldots, l_1)$ and partial derivative values $x^{(j)}, \frac{\partial f}{\partial x_k}\Big|_{x=x^{(j)}} \in \mathbb{R}^n \times \mathbb{R}$, $k = 1, 2, \ldots, n; j = 1, 2, \ldots, l_2$, where x_k denotes the kth component of the vector x. The RLSVRD model (Jayadeva et al. [29, 30]) for the simultaneous learning of a real valued function and its derivatives in the n-dimensional case finds a regressor that estimates f and $\frac{\partial f}{\partial x_k}$, for $k = 1, 2, \ldots, n$.

To this end, it is considered that a mapping $\phi : \mathbb{R}^n \to \mathbb{R}^d$, $(d > n)$ exists such that

$$\phi(x^{(i)}) \equiv [\phi_1(x^{(i)}), \phi_2(x^{(i)}), \ldots, \phi_d(x^{(i)})]^T, (i = 1, 2, \ldots, l_1),$$

and

$$\phi^{'k}(x^{(j)}) \equiv [\frac{\partial \phi_1(x^{(j)})}{\partial x_k}, \frac{\partial \phi_2(x^{(j)})}{\partial x_k}, \ldots, \frac{\partial \phi_d(x^{(j)})}{\partial x_k}]$$

$k = 1, 2, \ldots, n; j = 1, 2, \ldots, l_2$.

The RLSVRD is then obtained by solving the following problem

$$\underset{(p,\, q^1,\, q^2, \ldots, q^n, w, b)}{\text{Min}} \quad \frac{1}{2}(w^T w + b^2) + \frac{c_1}{2} p^T p + \sum_{k=1}^{n} \frac{c_2{}^k}{2} q^{k^T} q^k$$

subject to

$$\begin{aligned} & w^T \phi(x^{(i)}) + b - y_i + p_i = 0, \quad (i = 1, 2, \ldots, l_1), \\ & w^T \phi^{'k}(x^{(i)}) - y_j^{'k} + q_j{}^k = 0, \; (j = 1, 2, \ldots, l_2, k = 1, 2, \ldots, n), \end{aligned} \tag{4.66}$$

where $c_1 > 0, c_2{}^k > 0, k = 1, 2, \ldots, n$ are parameters. The solution can be obtained by solving the following set of equations

$$\begin{bmatrix} HH^T + ee^T + I/c_1 & HG_1{}^T & \ldots & HG_n{}^T \\ G_1 H^T & G_1 G_1{}^T + I/c_2{}^1 & \ldots & G_1 G_n{}^T \\ \vdots & \vdots & \ddots & \vdots \\ G_n H^T & \ldots & \ldots & G_n G_n{}^T + I/c_2{}^n \end{bmatrix} \begin{bmatrix} \alpha \\ \beta^1 \\ \beta_2 \\ \vdots \\ \beta^n \end{bmatrix} = \begin{bmatrix} Y \\ Y^{'1} \\ Y^{'2} \\ \vdots \\ Y^{'n} \end{bmatrix},$$

where, $Z = [x^{(1)}, x^{(2)}, \ldots, x^{(l_1)}]^T$, $Z_1 = [x^{(1)}, x^{(2)}, \ldots, x^{(l_2)}]^T$, $Y^{'k} = [\frac{\partial f}{\partial x_k}]|_{x = x^{(j)}}$, $H = [\phi(Z)]_{l_1 \times d}$ and $G_k = [\phi^{'k}(Z_1)]_{l_2 \times d}$, $(k = 1, 2, \ldots, n, j = 1, 2, \ldots, l_2)$. Once α and β^k, $(k = 1, 2, \ldots, n)$ are obtained, the estimated values of the function and its derivatives at any $x \in \mathbb{R}^n$ are then given by

$$f(x) = \sum_{i=1}^{l_1} \alpha_i \phi^T(x^{(i)}) \phi(x) + \sum_{k=1}^{n} \sum_{j=1}^{l_2} \beta_j{}^k [\phi^{'k}(x^{(j)})]^T \phi(x) + e^T \alpha,$$

and

$$f^{'k}(x) = \sum_{i=1}^{m_1} \alpha_i \phi^T(x^{(i)}) [\phi^{'k}(x)] + \sum_{j=1}^{m_2} \beta_j{}^k [\phi^{'k}(x^{(j)})]^T [\phi^{'k}(x)], \quad (k = 1, 2, \ldots, n).$$

We note that although RLSVRD model improves over Lazaro et al. model in terms of estimation accuracy and computational complexity but its least square approach lacks sparseness making it unfit for large datasets.

4.5.2 TSVRD: Twin Support Vector Regression of a Function and Its Derivatives

Here we describe the TSVRD model of Khemchandani et al. [31] for simultaneously learning a function and its derivatives. For a real valued function $f : \mathbb{R}^n \rightarrow \mathbb{R}$, let A of dimension $(l_1 \times n)$ be the set of points $\{x^{(i)} \in \mathbb{R}^n,\ i = 1, 2, \ldots, l_1\}$ over which functional values y are provided and B_j of dimension $(l_{2j} \times n)$ be the set of points over which partial derivatives $\frac{\partial f}{\partial x_j}$ are provided, $j = 1, 2, \ldots, n$. Here x_j denote the jth component of the vector x.

Let $B = \begin{bmatrix} B_1 \\ B_2 \\ \vdots \\ B_n \end{bmatrix}$ (size: $l_2 \times n$ where $l_2 = (l_{21} + l_{22} + \cdots + l_{2n})$) and let $C = \begin{bmatrix} A \\ B \end{bmatrix}$ (size: $(l_1 + l_2) \times n$). Let Y be the set containing the values of the function at A, and $Y_j^{'}$ be the set of jth derivative values of the function prescribed at B_j. Let $Y^{'} = \begin{bmatrix} Y_1^{'} \\ Y_2^{'} \\ \vdots \\ Y_n^{'} \end{bmatrix}$, $Z = \begin{bmatrix} Y \\ Y^{'} \end{bmatrix}$.

On the similar lines to TSVR (Peng [3]), ϵ_1 and ϵ_2 perturbations in Z are introduced in order to create the upper-bound and lower-bound regression tubes for the simultaneous learning of a function and its derivatives, where $\epsilon_1 = \begin{bmatrix} \epsilon_{11} \\ \epsilon_{12} \end{bmatrix}$ and $\epsilon_2 = \begin{bmatrix} \epsilon_{21} \\ \epsilon_{22} \end{bmatrix}$. Y and $Y_j^{'}, j = 1, 2, \ldots, n$, are then regressed according to the following functions

$$Y - e'\epsilon_{11} = K_{\sigma_1}(A, C^T)w_1 + eb_1, \tag{4.67}$$

$$Y_j^{'} - e'\epsilon_{12} = K^{'}_{\sigma_{1j}x_j}(B_j, C^T)w_1, \quad (j = 1, 2, \ldots, n), \tag{4.68}$$

and

$$Y + e'\epsilon_{21} = K_{\sigma_2}(A, C^T)w_2 + eb_2, \tag{4.69}$$

$$Y_j^{'} + e'\epsilon_{22} = K^{'}_{\sigma_{2j}x_j}(B_j, C^T)w_2, \quad (j = 1, 2, \ldots, n), \tag{4.70}$$

where (w_1, b_1) and (w_2, b_2) define the upper-tube and lower-tube regressions of the function and its derivatives, $K(x, y)$ is the kernel used with $K^{'}_{x_j}(x, y)$ as the derivative of $K(x, y)$ with respect to $x_j, j = 1, 2, \ldots, n$ and e' is a vector of ones of appropriate dimension.

Let $e = \begin{bmatrix} e_1 \\ e_2 \end{bmatrix}$ be a column vector of length $(l_1 + l_2)$, e_1 be a column vector of ones of length l_1 and o_1 be a column vector of zeros of length l_2. Let $e^* = \begin{bmatrix} e_1 \\ o_1 \end{bmatrix}$,
$G^* = \begin{bmatrix} K_{\sigma_1}(A, C^T) \\ K'_{\sigma_{11}x_1}(B_1, C^T) \\ \vdots \\ K'_{\sigma_{1n}x_n}(B_n, C^T) \end{bmatrix}$, $u_i = \begin{bmatrix} w_i \\ b_i \end{bmatrix}$, $(i = 1, 2)$ and $G = [G^* \quad e^*]$.

Then the two quadratic programming optimization problems(QPPs) for the upper-bound and lower-bound regression functions respectively are

$$\text{(TSVRD1)} \quad \underset{(u_1,\ \xi_1)}{\text{Min}} \quad \frac{1}{2}\|(Z - e\epsilon_1 - Gu_1)\|^2 + \frac{1}{2}C_1({u_1}^T u_1) + C_2 e^T \xi_1$$
$$\text{subject to}$$
$$Z - e\epsilon_1 - Gu_1 \geq -\xi_1,$$
$$\xi_1 \geq 0. \tag{4.71}$$

and

$$\text{(TSVRD2)} \quad \underset{(u_2,\ \xi_2)}{\text{Min}} \quad \frac{1}{2}\|(Z + e\epsilon_2 - Gu_2)\|^2 + \frac{1}{2}C_3({u_2}^T u_2) + C_4 e^T \xi_2$$
$$\text{subject to}$$
$$Gu_2 - (Z + e\epsilon_2) \geq -\xi_2,$$
$$\xi_2 \geq 0, \tag{4.72}$$

where $Gu_1 = [G^* \quad e^*]\begin{bmatrix} w_1 \\ b_1 \end{bmatrix}$, $Gu_2 = [G^* \quad e^*]\begin{bmatrix} w_2 \\ b_2 \end{bmatrix}$, C_1, C_2, C_3, C_4 are the regularization parameters, and ξ_1, ξ_2 are non-negative error variables.

The above two QPP solves for u_1 and u_2 respectively finding the bounding regressors parameters (w_1, b_1) and (w_2, b_2). The final regressor of the function (and its derivatives) are then given by the mean of the upper and lower bound regressor tubes of the function (and its derivatives).

$$f(x) = \frac{1}{2}(K_{\sigma_1}(x, C^T)w_1 + K_{\sigma_2}(x, C^T)w_2) + \frac{1}{2}(b_1 + b_2) \tag{4.73}$$
$$\frac{\partial f}{\partial x}|_{x=x_j} = \frac{1}{2}(K'_{\sigma_{1j}x_j}(x, C^T)w_1 + K'_{\sigma_{2j}x_j}(x, C^T)w_2), \quad j = 1, 2 \ldots n. \tag{4.74}$$

4.5.3 GEPSVRD: Generalized Eigenvalue Proximal Support Vector Regression of a Function and Its Derivatives

In this section, similar to the lines of GEPSVR (Khemchandani et al. [32]), we discuss the GEPSVR formulation of Goyal [16] for simultaneous learning a function and its derivatives(GEPSVRD) which requires solving two generalized eigenvalue problems. The objective of the GEPSVRD algorithm is to find two non parallel ϵ-insensitive bounding regressors for functions and its derivatives. We first present the approach for a real valued function of a single variable f: $\mathbb{R} \rightarrow \mathbb{R}$ and later extend the same to functions of n variables.

The GEPSVRD algorithm finds regressors of the function, $f_1(x) = K_{\sigma_1}(x, C^T) w_1 + b_1$ and $f_2(x) = K_{\sigma_2}(x, C^T)w_2 + b_2$ and regressors of the derivative function, $f_1'(x) = K'_{\sigma_1'}(x, C^T)w_1$ and $f_2'(x) = K'_{\sigma_2'}(x, C^T)w_2$ corresponding to the ϵ-insensitive bounding regressor of the function and its derivative. Here, $K(x, y) = \phi(x)^T\phi(y)$ is the kernel employed and $K'(x, y)$ is its derivative with respect to x. The end regressor of the function (and its derivative) is then given by the mean of the functions $f_1(x)$ and $f_2(x)$ (respectively $f_1'(x)$ and $f_2'(x)$).

Let A of dimension $(l_1 \times 1)$ be the set of points over which functional values y are provided and B of dimension $(l_2 \times 1)$ be the set of points over which derivative values are provided, and let $C = \begin{bmatrix} A \\ B \end{bmatrix}$ (size: $(l_1 + l_2) \times 1$). Let Y be the set containing the values of the function at A, and $Y^{'}$ be the set of derivative values of the function prescribed at B. Let $Z = \begin{bmatrix} Y \\ Y^{'} \end{bmatrix}$.

Similar to the spirit of GEPSVR, the GEPSVRD formulations aims to determine $f_1(x)$ (and $f_1'(x)$) so as to minimize the Euclidean distance between the plane from the set of points $(K(A, C), Y - \epsilon_{11}e')$ (respectively $(K(B, C), Y' - \epsilon_{12}e')$) (training points translated downwards by ϵ_{11} (respectively ϵ_{12})) and maximize its Euclidean distance from the set of points $(K(A, C), Y + \epsilon_{21}e')$ (respectively $(K(B, C), Y' + \epsilon_{22}e')$) (training points translated upwards by ϵ_{21} (respectively ϵ_{22})). Similarly, $f_2(x)$ (and $f_2'(x)$) is determined so as to minimize the Euclidean distance between the plane from the set of points $(K(A, C), Y + \epsilon_{21}e')$ (respectively $(K(B, C), Y' + \epsilon_{22}e')$) and maximize its Euclidean distance from the set of points $(K(A, C), Y - \epsilon_{11}e')$ (respectively $(K(B, C), Y' - \epsilon_{12}e')$), where e' is a vector of appropriate dimension.

This results in the following optimization problems

$$\underset{u_1 \neq 0}{\text{Min}} \quad \frac{\|(Gu_1 - (Z - e\epsilon_1))\|^2/\|u_1\|^2}{\|(Gu_1 - (Z + e\epsilon_2))\|^2/\|u_1\|^2}, \tag{4.75}$$

and

$$\underset{u_2 \neq 0}{\text{Min}} \quad \frac{\|(Gu_2 - (Z + e\epsilon_2))\|^2/\|u_2\|^2}{\|(Gu_2 - (Z - e\epsilon_1))\|^2/\|u_2\|^2}, \tag{4.76}$$

where $\epsilon_1 = \begin{bmatrix} \epsilon_{11} \\ \epsilon_{12} \end{bmatrix}, \epsilon_2 = \begin{bmatrix} \epsilon_{21} \\ \epsilon_{22} \end{bmatrix}, e = \begin{bmatrix} e_1 \\ e_2 \end{bmatrix}$ be a column vector of length $(l_1 + l_2)$, e_1 be a column vector of ones of length l_1 and o_1 be a column vector of zeros of length l_2, $e^* = \begin{bmatrix} e_1 \\ o_1 \end{bmatrix}, G^* = \begin{bmatrix} K_{\sigma_1}(A, C^T) \\ K'_{\sigma'_1}(B, C^T) \end{bmatrix}, u_i = \begin{bmatrix} w_i \\ b_i \end{bmatrix}, i = 1, 2$ and $G = [G^* \quad e^*]$.

Here it is implicitly assumed that $u_i \neq 0,\ i = 1, 2$ implies that $(Gu_1 - (Z - e\epsilon_1)) \neq 0$ and $(Gu_2 - (Z + e\epsilon_2)) \neq 0$ (Khemchandani et al. [32]; Mangasarian and Wild [33]). In that case (4.75) can be simplified as

$$\underset{u_1 \neq 0}{\text{Min}} \quad \frac{\|(Gu_1 - (Z - e\epsilon_1))\|^2}{\|(Gu_1 - (Z + e\epsilon_2))\|^2}. \tag{4.77}$$

Now the optimization problem (4.77) can be regularized by introducing Tikhonov regularization term (Tikhonov and Arsenin [34], Saunders et al. [21]) as

$$\underset{u_1 \neq 0}{\text{Min}} \quad \frac{\|(Gu_1 - (Z - e\epsilon_1))\|^2 + \delta_1 \|[u_1 \quad -1]^T\|^2}{\|(Gu_1 - (Z + e\epsilon_2))\|^2}, \tag{4.78}$$

where $\delta_1 > 0$ is the regularization coefficient. Let $\nu_1 = [u_1 \quad -1]^T$ and hence the above problem converts to Rayleigh Quotient of the form

$$\underset{\nu_1 \neq 0}{\text{Min}} \quad \frac{\nu_1^T R \nu_1}{\nu_1^T S \nu_1}, \tag{4.79}$$

where

$$\begin{aligned} R &= [G \quad (Z - e\epsilon_1)]^T [G \quad (Z - e\epsilon_1)] + \delta_1 I, \\ S &= [G \quad (Z + e\epsilon_2)]^T [G \quad (Z + e\epsilon_2)]. \end{aligned} \tag{4.80}$$

Using Rayleigh Quotient properties (Mangasarian and Wild [33]; Parlett [35]), the solution of (4.79) is obtained by solving the following generalized eigenvalue problem

$$R\nu_1 = \eta_1 S \nu_1, \quad \nu_1 \neq 0. \tag{4.81}$$

Let μ_1 denote the eigenvector corresponding to the smallest eigenvalue η_{min} of (4.81). To obtain u_1 from μ_1, we normalize μ_1 by the negative of the $(l_1 + l_2 + 2)^{th}$ element of μ_1 so as to force a (-1) at the $(l_1 + l_2 + 2)^{th}$ position of μ_1. Let this normalized representation of μ_1 be μ'_1 with (-1) at the end, such that $\mu'_1 = [u_1 \quad -1]^T$. We get w_1 and b_1 from u_1 which determines an ϵ-insensitive bounding regressors $f_1(x) = K_{\sigma_1}(x, C^T)w_1 + b_1$ and $f'_1(x) = K'_{\sigma'_1}(x, C^T)w_1$. We note that the solution is easily obtained using a single MATLAB command that solves the classical generalized eigenvalue problem.

In a similar manner, $f_2(x)$ and $f_2'(x)$ is determined by considering the optimization problem (4.76) along with the Tikhonov regularization as follows

$$\underset{u_2 \neq 0}{\text{Min}} \quad \frac{\|(Gu_2 - (Z + e\epsilon_2))\|^2 + \delta_2 \|[u_2 \quad -1]^T\|^2}{\|(Gu_2 - (Z - e\epsilon_1))\|^2}, \tag{4.82}$$

where $\delta_2 > 0$ is is the regularization constant. Let $v_2 = [u_2 \quad -1]^T$ and hence the above problem converts to Rayleigh Quotient of the form

$$\underset{v_2 \neq 0}{\text{Min}} \quad \frac{v_2^T P v_2}{v_2^T Q v_2}, \tag{4.83}$$

where

$$\begin{aligned} P &= [G \quad (Z + e\epsilon_2)]^T [G \quad (Z + e\epsilon_2)] + \delta_2 I, \\ Q &= [G \quad (Z - e\epsilon_1)]^T [G \quad (Z - e\epsilon_1)]. \end{aligned} \tag{4.84}$$

The solution of (4.83) is thus obtained by solving the following generalized eigenvalue problem

$$P v_2 = \eta_2 Q v_2, \quad v_2 \neq 0. \tag{4.85}$$

Finding the minimum eigenvalue of (4.85) and then normalizing the corresponding eigenvector in the similar manner mentioned for problem (4.81), we get u_2 from which w_2 and b_2 is obtained determining an ϵ-insensitive bounding regressors $f_2(x) = K_{\sigma_2}(x, C^T)w_2 + b_2$ and $f_2'(x) = K'_{\sigma_2'}(x, C^T)w_2$.

We next extend the formulations to consider the real valued functions of n variables $f : \mathbb{R}^n \to \mathbb{R}$. Let A of dimension $(l_1 \times n)$ be the set of points $\{x^{(i)} \in \mathbb{R}^n, i = 1, 2, \ldots, l_1\}$ over which functional values y are provided and B_j of dimension $(l_{2j} \times n)$ be the set of points over which partial derivatives $\frac{\partial f}{\partial x_j}$ are provided, $j = 1, 2, \ldots, n$. Here x_j denote the jth component of the vector x. Let $B = \begin{bmatrix} B_1 \\ B_2 \\ \vdots \\ B_n \end{bmatrix}$ (size: $l_2 \times n$ where $l_2 = (l_{21} + l_{22} + \cdots + l_{2n})$) and let $C = \begin{bmatrix} A \\ B \end{bmatrix}$ (size: $(l_1 + l_2) \times n$). Let Y be the set containing the values of the function at A, and Y_j' be the set of jth derivative values of the function prescribed at B_j. Let $Y' = \begin{bmatrix} Y_1' \\ Y_2' \\ \vdots \\ Y_n' \end{bmatrix}$, $Z = \begin{bmatrix} Y \\ Y' \end{bmatrix}$.

The algorithm finds regressors of the function, $f_1(x) = K_{\sigma_1}(x, C^T)w_1 + b_1$ and $f_2(x) = K_{\sigma_2}(x, C^T)w_2 + b_2$ and regressors of the derivative function, $f'_{1j}(x) = K'_{\sigma'_1 x_j}(x, C^T)w_1$ and $f'_{2j}(x) = K'_{\sigma'_2 x_j}(x, C^T)w_2$ corresponding to the ϵ-insensitive bounding regressor of the function and its partial derivatives, where $K'_{x_j}(x, y)$ is the derivative of $K(x, y)$ with respect jth dimension of input variable $x_j, j = 1, 2, \ldots, n$.

Let $e = \begin{bmatrix} e_1 \\ e_2 \end{bmatrix}$ be a column vector of length $(l_1 + l_2)$, e_1 be a column vector of ones of length l_1 and o_1 be a column vector of zeros of length l_2. Let $e^* = \begin{bmatrix} e_1 \\ o_1 \end{bmatrix}$,

$$G^* = \begin{bmatrix} K_{\sigma_1}(A, C^T) \\ K'_{\sigma_{11} x_1}(B_1, C^T) \\ \vdots \\ K'_{\sigma_{1n} x_n}(B_n, C^T) \end{bmatrix}, u_i = \begin{bmatrix} w_i \\ b_i \end{bmatrix}, i = 1, 2 \text{ and } G = [G^* \quad e^*].$$

Similar to the case for single variable, the two optimization problems are

$$\underset{u_1 \neq 0}{\text{Min}} \frac{\|(Gu_1 - (Z - e\epsilon_1))\|^2 / \|u_1\|^2}{\|(Gu_1 - (Z + e\epsilon_2))\|^2 / \|u_1\|^2}, \tag{4.86}$$

and

$$\underset{u_2 \neq 0}{\text{Min}} \frac{\|(Gu_2 - (Z + e\epsilon_2))\|^2 / \|u_2\|^2}{\|(Gu_2 - (Z - e\epsilon_1))\|^2 / \|u_2\|^2}. \tag{4.87}$$

We solve for u_1 and u_2 in a similar manner as in case of single variable functions obtaining (w_1, b_1) and (w_2, b_2).

The final regressors of the function (and its derivatives) are then given by the mean of the upper and lower bound regressor tubes of the function (and its derivatives).

$$f(x) = \frac{1}{2}(K_{\sigma_1}(x, C^T)w_1 + K_{\sigma_2}(x, C^T)w_2) + \frac{1}{2}(b_1 + b_2) \tag{4.88}$$

$$\frac{\partial f}{\partial x}|_{x=x_j} = \frac{1}{2}(K'_{\sigma_{1j} x_j}(x, C^T)w_1 + K'_{\sigma_{2j} x_j}(x, C^T)w_2), \quad j = 1, 2 \ldots n. \tag{4.89}$$

4.5.4 Regularized GEPSVRD

In this subsection, we present the regularization technique discussed in Khemchandani et al. [32] and Guarracino et al. [36] for finding a single eigenvalue problem whose eigenvectors corresponding to minimum and maximum eigenvalue would provide the parameter values for both the ϵ-insensitive regressors ($u_i = \begin{bmatrix} w_i \\ b_i \end{bmatrix}$, $i = 1, 2$) of function and its derivative. We consider the following regularized optimization problem

$$\underset{u \neq 0}{\text{Min}} \frac{\|(Gu - (Z - e\epsilon_1))\|^2 + \hat{\delta_1}\left\|(\hat{G}u - (Z + e\epsilon_2))\right\|^2}{\|(Gu - (Z + e\epsilon_2))\|^2 + \hat{\delta_2}\left\|(\hat{G}u - (Z - e\epsilon_1))\right\|^2}. \tag{4.90}$$

where $\hat{\delta_1}$, $\hat{\delta_2}$ are non-negative regularization constant and $\hat{G} = [\hat{G^*} \ e^*]$ where $\hat{G^*}$ is the diagonal matrix of diagonal values of G^*. From the work of Guarracino et al. [36], we get that the minimum eigenvalue of the original problem (4.77) becomes the maximum eigenvalue of the above problem and the maximum eigenvalue of the original problem (4.77) becomes the minimum eigenvalue of the above problem. Let $t = [u \ \ -1]^T$ and U and V be defined as

$$U = [G \quad (Z - e\epsilon_2)]^T[G \quad (Z - e\epsilon_2)] + \hat{\delta_1}[\hat{G} \quad (Z + e\epsilon_1)]^T[\hat{G}(Z + e\epsilon_1)],$$
$$V = [G \quad (Z + e\epsilon_1)]^T[G \quad (Z + e\epsilon_1)] + \hat{\delta_2}[\hat{G} \quad (Z - e\epsilon_2)]^T[\hat{G} \quad (Z - e\epsilon_2)].$$

Now again using the properties of Rayleigh quotient, the optimization problem (4.90) is equivalent to following generalized eigenvalue problem

$$Ut = \nu Vt \quad , \quad t \neq 0. \tag{4.91}$$

Let t_1 denotes the eigenvector corresponding to the maximum eigenvalue of (4.91) and t_2 corresponds to the minimum eigenvalue of (4.91). To obtain u_1 and u_2 from t_1 and t_2, we follow similar normalization procedure of Sect. 4.5.3 and get $t_1' = (u_1 \ \ -1)$ and $t_2' = (u_2 \ \ -1)$. From this, we get w_1, w_2 and b_1, b_2 and hence both the ϵ-insensitive regressors of the function and its derivatives are determined. The final regressors of the function and its derivatives are determined in the same way by calculating the mean of the bounding regressors.

We note that using this regularized technique, we only need to solve one generalized eigenvalue problem to obtain the final regressor as opposed to solving two generalized eigenvalue problems. This results in similar performance in terms of estimation accuracy but provides the advantage of lower computational complexity.

4.5.5 Experimental Results

To prove the efficacy of the proposed GEPSVRD approach over above mentioned approaches, comparison of estimation accuracy and run time complexity have been performed on $sinc(x)$, $xcos(x)$, $xsin(y)$, and seven functions of two real variables which were introduced in Lázarao et al. [27]. All the methods have been implemented in MATLAB 7.8 running on a PC with Intel Core 2 Duo processor (2.00 GHz) with 3 GB of RAM. Here, we note that the time complexity of Lazaro et al. approach is shown to be higher than the RLSVRD and TSVRD approaches (Jayadeva et al. [29, 30]; Khemchandani et al. [31]) and therefore comparison of only the run time of discussed model GEPSVRD (Goyal [16]) with RLSVRD and TSVRD approach.

The experimental setup is kept similar to the TSVRD experiment section (Khemchandani et al. [31]). For the sake of completion, we discuss the details here. We use five evaluation criteria to assess the performance of the above mentioned algorithms. Let y_i and $\hat{y}_i$, denote the original and predicted value of the func-

Table 4.5 Estimation accuracy comparisons for sinc(x), xcos(x) and xsin(y)

		RAE	MAE	RMSE	SSE/SST	SSR/SST
sinc(x)						
GEPSVRD	$f(x)$	0.0192	0.00002	0.00004	3×10^{-8}	1.00002
	$f'(x)$	0.0457	0.0001	0.0002	2×10^{-7}	1.00003
TSVRD	$f(x)$	0.6423	0.0009	0.0013	0.00002	0.9968
	$f'(x)$	0.5985	0.0019	0.0027	0.00003	0.9943
RLSVRD	$f(x)$	2.0192	0.0028	0.0040	0.0002	0.9708
	$f'(x)$	2.2825	0.0073	0.0091	0.0004	0.9648
SVRD (Lazaro et al.)	$f(x)$	1.7577	0.0024	0.0043	0.00028	0.9685
	$f'(x)$	1.7984	0.0057	0.0077	0.00026	0.9683
xcos(x)						
GEPSVRD	$f(x)$	0.0888	0.0022	0.0039	0.000001	0.9999
	$f'(x)$	0.3905	0.0087	0.0129	0.000024	0.9998
TSVRD	$f(x)$	0.1142	0.0029	0.0049	0.000002	1.0004
	$f'(x)$	0.4900	0.0109	0.0172	0.00004	1.0005
RLSVRD	$f(x)$	0.3540	0.0088	0.0134	0.0000	0.9925
	$f'(x)$	0.5510	0.0123	0.0183	0.0000	0.9940
SVRD (Lazaro et al.)	$f(x)$	2.8121	0.0702	0.0941	0.0009	0.9430
	$f'(x)$	2.3863	0.0531	0.0655	0.0006	0.9559
xsin(y)						
GEPSVRD	$f(x)$	0.5723	0.0056	0.0085	0.00004	1.0028
	$\frac{\partial f}{\partial x}\|_{x=x_1}$	0.6789	0.0041	0.0056	0.00006	0.9932
	$\frac{\partial f}{\partial x}\|_{x=x_2}$	0.4103	0.0043	0.0071	0.00003	0.9990
TSVRD	$f(x)$	0.8242	0.0081	0.0122	0.0001	1.0053
	$\frac{\partial f}{\partial x}\|_{x=x_1}$	1.1843	0.0072	0.0101	0.0002	0.9986
	$\frac{\partial f}{\partial x}\|_{x=x_2}$	0.6730	0.0070	0.0118	0.0001	1.0018
RLSVRD	$f(x)$	0.8544	0.0084	0.0130	0.0001	0.9873
	$\frac{\partial f}{\partial x}\|_{x=x_1}$	2.9427	0.0180	0.0269	0.0015	0.9640
	$\frac{\partial f}{\partial x}\|_{x=x_2}$	1.2276	0.0128	0.0211	0.0002	0.9795
SVRD (Lazaro et al.)	$f(x)$	1.0094	0.0100	0.0140	0.0001	0.9865
	$\frac{\partial f}{\partial x}\|_{x=x_1}$	1.6307	0.0100	0.0131	0.0004	0.9765
	$\frac{\partial f}{\partial x}\|_{x=x_2}$	1.5325	0.0160	0.0236	0.0003	0.9760

tion/derivatives respectively at a training point x_i. Let m be the size of the testing dataset and let $\bar{y} = \sum_{i=1}^{l} y_i$. The five evaluation measures are then defined as:

1. Relative Absolute Error (RAE):

$$\text{RAE} = \frac{\sum_{i=1}^{l} |y_i - \hat{y}_i|}{\sum_{i=1}^{l} |y_i|} \times 100.$$

2. Mean Absolute Error (MAE):

$$\text{MAE} = \frac{1}{l}\sum_{i=1}^{l} |y_i - \hat{y}_i|.$$

3. Root Mean Squared Error (RMSE):

$$\text{RMSE} = \sqrt{\frac{1}{l}\sum_{i=1}^{l}(y_i - \hat{y}_i)^2}.$$

4. SSE/SST, which is defined as the ratio between the Sum Squared Error ($\text{SSE} = \sum_{i=1}^{l}(y_i - \hat{y}_i)^2$) and Sum Squared Deviation of testing samples ($\text{SST} = \sum_{i=1}^{l} (y_i - \bar{y})^2$).
5. SSR/SST, which is defined as the ratio between the Interpretable Sum Squared Deviation ($\text{SSR} = \sum_{i=1}^{l}(\hat{y}_i - \bar{y})^2$) and Sum Squared Deviation of testing samples (SST).

In most cases, a lower value of RAE, MAE, RMSE and SSE/SST reflects precision in agreement between the estimated and original values, while a higher value of SSR/SST shows higher statistical information being accounted by the regressor.

Table 4.6 Functions of 2 variables provided by Lazaro et al.

Name	Function	Domain
Function 1	$y = \sin(x_1, x_2)$	$[-2, 2]$
Function 2	$y = \exp(x_1 \sin(\pi x_2))$	$[-1, 1]$
Function 3	$y = \frac{40\times\exp(8(x_1-0.5)^2)+(x_2-0.5)^2}{\exp(8((x_1-0.2)^2+(x_2-0.7)^2))+\exp(8((x_1-0.7)^2+(x_2-0.2)^2))}$	$[0, 1]$
Function 4	$y = (1 + \sin(2x_1 + 3x_2))/(3.5 + sin(x_1 - x_2))$	$[-2, 2]$
Function 5	$y = 42.659(0.1 + x_1(0.05 + {x_1}^4 - 10{x_1}^2{x_2}^2 + 5{x_2}^4))$	$[-0.5, 0.5]$
Function 6	$y =$ $1.3356(\exp(3(x_2 - 0.5))\sin(4\pi(x_2 - 0.9)^2 + 1.5(1 - x_1))$ $+1.5(1 - x_1)) + \exp(2x_1 - 1)sin(3\pi(x_1 - 0.6)^2)$	$[0, 1]$
Function 7	$y = 1.9[1.35 + \exp(x_1)\sin(13(x_1 - 0.6)^2)$ $+\exp(3(x_2 - 0.5))\sin(4\pi(x_2 - 0.9)^2)]$	$[0, 1]$

Table 4.7 Estimation accuracy comparisons for functions 1–2

		RAE	MAE	RMSE	SSE/SST	SSR/SST	Time (in sec)
Function 1							
GEPSVRD	$f(x)$	0.3439	0.0018	0.0025	0.00002	1.0005	0.661
	$\frac{\partial f}{\partial x}\lvert_{x=x_1}$	1.3178	0.0091	0.0119	0.00018	1.0002	
	$\frac{\partial f}{\partial x}\lvert_{x=x_2}$	1.4887	0.0103	0.0147	0.0003	1.0010	
TSVRD	$f(x)$	0.5308	0.0029	0.0036	0.0000	1.0011	2.949
	$\frac{\partial f}{\partial x}\lvert_{x=x_1}$	2.0930	0.0145	0.0245	0.0008	0.9927	
	$\frac{\partial f}{\partial x}\lvert_{x=x_2}$	2.4760	0.0171	0.0283	0.0010	0.9933	
RLSVRD	$f(x)$	2.2207	0.0120	0.0175	0.0008	0.9632	0.088
	$\frac{\partial f}{\partial x}\lvert_{x=x_1}$	5.1797	0.0358	0.0607	0.0048	0.9162	
	$\frac{\partial f}{\partial x}\lvert_{x=x_2}$	5.0381	0.0348	0.0642	0.0054	0.9294	
SVRD (Lazaro et al.)	$f(x)$	3.0131	0.0163	0.0202	0.0010	0.9443	
	$\frac{\partial f}{\partial x}\lvert_{x=x_1}$	5.2745	0.0365	0.0607	0.0048	0.9143	
	$\frac{\partial f}{\partial x}\lvert_{x=x_2}$	5.2744	0.0365	0.0607	0.0048	0.9143	
Function 2							
GEPSVRD	$f(x)$	0.2164	0.0023	0.0038	0.00001	1.0005	0.627
	$\frac{\partial f}{\partial x}\lvert_{x=x_1}$	2.8985	0.0196	0.0286	0.0010	1.0094	
	$\frac{\partial f}{\partial x}\lvert_{x=x_2}$	1.0361	0.0123	0.0192	0.0001	1.0054	
TSVRD	$f(x)$	0.4298	0.0047	0.0075	0.0003	1.0123	2.393
	$\frac{\partial f}{\partial x}\lvert_{x=x_1}$	3.9749	0.0268	0.0565	0.0040	1.0293	
	$\frac{\partial f}{\partial x}\lvert_{x=x_2}$	2.0856	0.0247	0.0430	0.0007	1.0123	
RLSVRD	$f(x)$	1.2459	0.0136	0.0188	0.0016	0.9531	0.1035
	$\frac{\partial f}{\partial x}\lvert_{x=x_1}$	9.0260	0.0609	0.1040	0.0137	0.9862	
	$\frac{\partial f}{\partial x}\lvert_{x=x_2}$	3.7336	0.0443	0.0667	0.0017	0.9775	
SVRD (Lazaro et al.)	$f(x)$	2.7539	0.0301	0.0341	0.0052	0.9527	
	$\frac{\partial f}{\partial x}\lvert_{x=x_1}$	6.6438	0.0448	0.0808	0.0083	0.9351	
	$\frac{\partial f}{\partial x}\lvert_{x=x_2}$	6.4934	0.0771	0.1278	0.0063	0.9258	

For our simulations of GEPSVRD, we have considered Gaussian Kernel and the values of parameters like sigma(σ), delta (δ) are selected by utilizing a cross validation approach. Both ϵ_1 and ϵ_2 are taken as 0.1 in all our simulations. We compare our method with TSVRD, RLSVRD and Lazaro's approach for which the optimal set of parameters have already been prescribed in Khemchandani et al. [31], Jayadeva et al. [29] and Lázarao et al. [27].

Table 4.8 Estimation accuracy comparisons for functions 3–4

		RAE	MAE	RMSE	SSE/SST	SSR/SST	Time (in sec)
Function 3							
GEPSVRD	$f(x)$	0.1659	0.0069	0.0086	0.000007	1.0001	0.654
	$\frac{\partial f}{\partial x}\|_{x=x_1}$	1.5048	0.1406	0.1868	0.00028	0.9997	
	$\frac{\partial f}{\partial x}\|_{x=x_2}$	1.5048	0.1406	0.1868	0.00028	0.9997	
TSVRD	$f(x)$	0.1717	0.0071	0.0094	0.0000	1.0000	1.8576
	$\frac{\partial f}{\partial x}\|_{x=x_1}$	1.5303	0.1429	0.1981	0.0003	0.9980	
	$\frac{\partial f}{\partial x}\|_{x=x_2}$	1.5303	0.1429	0.1981	0.0003	0.9980	
RLSVRD	$f(x)$	0.6790	0.0282	0.0457	0.0002	0.9810	0.0769
	$\frac{\partial f}{\partial x}\|_{x=x_1}$	2.5077	0.2342	0.3395	0.0009	0.9848	
	$\frac{\partial f}{\partial x}\|_{x=x_2}$	2.5075	0.2342	0.3395	0.0009	0.9848	
SVRD (Lazaro et al.)	$f(x)$	26.3936	1.0944	1.3285	0.1675	0.7339	
	$\frac{\partial f}{\partial x}\|_{x=x_1}$	19.5264	1.8238	3.0800	0.0763	0.6923	
	$\frac{\partial f}{\partial x}\|_{x=x_2}$	19.5266	1.8238	3.0800	0.0763	0.6923	
Function 4							
GEPSVRD	$f(x)$	0.2249	0.0007	0.0010	0.00002	1.0013	0.667
	$\frac{\partial f}{\partial x}\|_{x=x_1}$	0.6247	0.0024	0.0033	0.00006	1.0006	
	$\frac{\partial f}{\partial x}\|_{x=x_2}$	0.4394	0.0025	0.0038	0.00003	1.0018	
TSVRD	$f(x)$	0.6813	0.0020	0.0032	0.0002	1.0026	2.5251
	$\frac{\partial f}{\partial x}\|_{x=x_1}$	1.7881	0.0068	0.0106	0.0006	1.0036	
	$\frac{\partial f}{\partial x}\|_{x=x_2}$	1.4216	0.0082	0.0134	0.0004	1.0065	
RLSVRD	$f(x)$	1.6019	0.0047	0.0063	0.0008	0.9648	0.0647
	$\frac{\partial f}{\partial x}\|_{x=x_1}$	3.0827	0.0117	0.0155	0.0013	0.9639	
	$\frac{\partial f}{\partial x}\|_{x=x_2}$	2.7079	0.0155	0.0202	0.0010	0.9699	
SVRD (Lazaro et al.)	$f(x)$	1.7264	0.0051	0.0063	0.0008	0.9657	
	$\frac{\partial f}{\partial x}\|_{x=x_1}$	2.0031	0.0076	0.0091	0.0004	0.9636	
	$\frac{\partial f}{\partial x}\|_{x=x_2}$	1.9077	0.0109	0.0131	0.0004	0.9646	

Table 4.9 Estimation accuracy comparisons for functions 5–7

		RAE	MAE	RMSE	SSE/SST	SSR/SST	Time (in sec)
Function 5							
GEPSVRD	$f(x)$	1.4484	0.0618	0.0778	0.0045	1.0372	0.661
	$\frac{\partial f}{\partial x}\|_{x=x_1}$	7.8831	0.6021	0.7644	0.0041	1.0213	
	$\frac{\partial f}{\partial x}\|_{x=x_2}$	14.5797	0.7799	0.9691	0.0147	1.0323	
TSVRD	$f(x)$	1.5022	0.0641	0.0805	0.0049	1.0324	1.0451
	$\frac{\partial f}{\partial x}\|_{x=x_1}$	8.3566	0.6382	0.8168	0.0047	1.0224	
	$\frac{\partial f}{\partial x}\|_{x=x_2}$	15.4407	0.8260	1.0297	0.0166	1.0177	
RLSVRD	$f(x)$	2.2546	0.0962	0.1194	0.0107	0.9429	0.1212
	$\frac{\partial f}{\partial x}\|_{x=x_1}$	8.9677	0.6848	0.9564	0.0064	0.9629	
	$\frac{\partial f}{\partial x}\|_{x=x_2}$	15.8932	0.8502	1.3877	0.0301	0.8361	
SVRD (Lazaro et al.)	$f(x)$	7.4122	0.3162	0.4083	0.1251	0.7302	
	$\frac{\partial f}{\partial x}\|_{x=x_1}$	31.2563	2.3869	5.6809	0.2274	0.4415	
	$\frac{\partial f}{\partial x}\|_{x=x_2}$	28.1296	1.5047	3.0378	0.1441	0.5786	
Function 6							
GEPSVRD	$f(x)$	0.7216	0.0159	0.0224	0.0004	0.9909	0.665
	$\frac{\partial f}{\partial x}\|_{x=x_1}$	4.7653	0.2334	0.3571	0.0037	0.9787	
	$\frac{\partial f}{\partial x}\|_{x=x_2}$	3.8634	0.2617	0.3221	0.0017	1.0035	
TSVRD	$f(x)$	0.8980	0.0198	0.0257	0.0006	0.9894	2.3506
	$\frac{\partial f}{\partial x}\|_{x=x_1}$	3.4963	0.1712	0.2893	0.0025	0.9794	
	$\frac{\partial f}{\partial x}\|_{x=x_2}$	5.7369	0.3887	0.4865	0.0040	0.9992	
RLSVRD	$f(x)$	2.3425	0.0518	0.0602	0.0035	0.9694	0.090
	$\frac{\partial f}{\partial x}\|_{x=x_1}$	3.1604	0.1548	0.2539	0.0019	0.9827	
	$\frac{\partial f}{\partial x}\|_{x=x_2}$	11.4298	0.7744	1.0357	0.0179	0.9312	
SVRD (Lazaro et al.)	$f(x)$	6.7198	0.1485	0.1777	0.0304	0.8634	
	$\frac{\partial f}{\partial x}\|_{x=x_1}$	9.7242	0.4762	0.8038	0.0191	0.8645	
	$\frac{\partial f}{\partial x}\|_{x=x_2}$	32.5400	2.2048	2.8528	0.1362	0.6473	
Function 7							
GEPSVRD	$f(x)$	0.4033	0.0109	0.0158	0.0002	1.0008	0.667
	$\frac{\partial f}{\partial x}\|_{x=x_1}$	5.5397	0.2529	0.3757	0.0034	0.9975	
	$\frac{\partial f}{\partial x}\|_{x=x_2}$	2.2828	0.1195	0.1855	0.0006	1.0008	
TSVRD	$f(x)$	0.5566	0.0150	0.0209	0.0004	1.0019	1.3801
	$\frac{\partial f}{\partial x}\|_{x=x_1}$	7.6981	0.3514	0.4857	0.0056	0.9986	
	$\frac{\partial f}{\partial x}\|_{x=x_2}$	2.9467	0.1542	0.2259	0.0009	1.0016	

(continued)

Table 4.9 (continued)

		RAE	MAE	RMSE	SSE/SST	SSR/SST	Time (in sec)
RLSVRD	$f(x)$	1.1144	0.0301	0.0385	0.0015	0.9946	0.0828
	$\frac{\partial f}{\partial x}\|_{x=x_1}$	12.7252	0.5809	0.8110	0.0156	0.9516	
	$\frac{\partial f}{\partial x}\|_{x=x_2}$	4.3384	0.2270	0.3574	0.0021	0.9812	
SVRD (Lazaro et al.)	$f(x)$	5.8414	0.1578	0.1734	0.0298	0.9343	
	$\frac{\partial f}{\partial x}\|_{x=x_1}$	16.2819	0.7432	1.2237	0.0356	0.8485	
	$\frac{\partial f}{\partial x}\|_{x=x_2}$	10.6524	0.5575	1.0347	0.0178	0.8602	

To compare the performance of GEPSVRD in the presence of noise in the data, we add a normally distributed noise $N(0, \delta)$ as introduced in Lázarao et al. [27, 28] to the function and derivative values while training.

We have first considered comparisons over the function $sinc(x) = \frac{sin(x)}{x}$ and $f(x) = xcos(x)$ on the basis of above mentioned performance evaluation criteria. We considered 47 and 36 points respectively over $[-7, 7]$ for both testing and training. The comparison results are provided in Table 4.5.

Table 4.5 also compares the performance of GEPSVRD over 2-variate function $f(x_1, x_2) = x_1 sin(x_2)$ involving two real-valued variables: x_1 and x_2 for which 169 points over $[-3, 3] \times [-3, 3]$ for both testing and training were taken.

We have also compared our method on seven functions, proposed by Lázarao et al. [27, 28], that are sampled over a two-dimensional input space. The analytical expression of each function is provided in Table 4.6. We have consider 121 points for training and 361 points for testing each of the seven functions. The comparison results are provided in Tables 4.7, 4.8 and 4.9.

We note the following from the experimental results

(i) The relative absolute error (RAE), mean absolute error (MAE), RMSE and SSE/SST are lower for GEPSVRD while estimating function and its derivatives as compared to other methods for almost all the datasets considered. This shows that our proposed method performs superiorly compared to other methods.
(ii) The SSR/SST is higher for GEPSVRD and TSVRD when compared to other methods indicating that GEPSVRD is also efficient algorithm for estimating function and its derivatives.
(iii) The run time of GEPSVRD is lower as compared to TSVRD. Since Khemchandani et al. [31] shows that run time complexity of TSVRD is lower than Lazaro's approach, therefore run time complexity of GEPSVRD model will also be lower compared to Lazaro's approach.

4.6 Conclusions

In this chapter, we review the formulation of Twin Support Vector Regression (TSVR) proposed by Peng [3] and conclude that this formulation is not in the true spirit of TWSVM. This is because Peng's formulation deals either with up or down bound regressor in each of the twin optimization problems. But our TWSVR model is in the true spirit of TWSVM where in each QPP, the objective function deals with up(down) bound regressor and constraints deal with down(up)-bound regressor. As a result of this strategy we solve two QPPs, each one being smaller than the QPP of classical SVR. The efficacy of the proposed algorithm has been established when compared with TSVR and SVR.

The proposed model of TWSVR is based on intuition, is shown to be mathematically derivable and follows the spirit of TWSVM making it not only better, but also correct choice for establishing future work on twin domain. Here it may be remarked that various existing extensions of Peng's model [3], e.g. Xu and Wang [4], Shao et al. [5], Chen et al. [6, 7], Zhao et al. [8], Zhang et al. [9], Peng [10, 11], Balasundaram and Tanveer [12] and Singh et al. [13], may be studied for our present model as well.

Another problem discussed in this chapter is the problem of simultaneous learning of function and its derivatives. Here after reviewing the work of Lázarao et al. [27, 28], formulation of the Twin Support Vector Regression of a Function and its Derivatives (TSVRD) due to Khemchandani et al. is discussed in detail. Further the formulation of Goyal [16] for a Generalized Eigenvalue Proximal Support Vector Regression of a Function and its Derivatives (GEPSVRD) is presented. The efficiency of TSVRD and GEPSVRD is demonstrated on various benchmark functions available in the literature.

References

1. Bi, J., & Bennett, K. P. (2003). A geometric approach to support vector regression. *Neurocomputing*, *55*, 79–108.
2. Jayadeva, Khemchandani, R., & Chandra, S. (2007). Twin support vector machines for pattern classification. *IEEE Transactions on Pattern Analysis and Machine Intelligence*, *29*(5), 905–910.
3. Peng, X. (2010). Tsvr: an efficient twin support vector machine for regression. *Neural Networks*, *23*, 365–372.
4. Xu, Y. T., & Wang, L. S. (2012). A weighted twin support vector regression. *Knowledge Based Systems*, *33*, 92–101.
5. Shao, Y. H., Zhang, C. H., Yang, Z. M., Żing, L., & Deng, N. Y. (2013). ϵ - Twin support vector machine for regression. *Neural Computing and Applications*, *23*(1), 175–185.
6. Chen, X., Yang, J., & Chen, L. (2014). An improved robust and sparse twin support vector regression via linear programming. *Soft Computing*, *18*, 2335–2348.
7. Chen, X. B., Yang, J., Liang, J., & Ye, Q. L. (2012). Smooth twin support vector regression. *Neural Computing and Applications*, *21*(3), 505–513.
8. Zhao, Y. P., Zhao, J., & Zhao, M. (2013). Twin least squares support vector regression. *Neurocomputing*, *118*, 225–236.

9. Zhang, P., Xu, Y. T., & Zhao, Y. H. (2012). Training twin support vector regression via linear programming. *Neural Computing and Applications*, *21*(2), 399–407.
10. Peng, X. (2012). Efficient twin parametric insensitive support vector regression. *Neurocomputing*, *79*, 26–38.
11. Peng, X. (2010). Primal twin support vector regression and its sparse approximation. *Neurocomputing*, *73*(16–18), 2846–2858.
12. Balasundaram, S., & Tanveer, M. (2013). On Lagrangian twin support vector regression. *Neural Computing and Applications*, *22*(1), 257–267.
13. Singh, M., Chadha, J., Ahuja, P., Jayadeva, & Chandra, S. (2011). Reduced twin support vector regression. *Neurocomputing*, *74*(9), 1471–1477.
14. Khemchandani, R., Goyal, K., & Chandra, S. (2015). Twin support vector machine based regression. *International Conference on Advances in Pattern Recognition*, 1–6.
15. Khemchandani, R., Goyal, K., & Chandra, S. (2015). TWSVR: regression via twin support vector machine. *Neural Networks*, *74*, 14–21.
16. Goyal, K. (2015) Twin Support Vector Machines Based Regression and its Extensions. M.Tech Thesis Report, Mathematics Department, Indian Institute of Technology, Delhi.
17. Deng, N., Tian, Y., & Zhang, C. (2012). *Support vector machines: Optimization based theory, algorithms and extensions*. New York: Chapman & Hall, CRC Press.
18. Shao, Y.-H., Zhang, C.-H., Wang, X.-B., & Deng, N.-Y. (2011). Improvements on twin support vector machines. *IEEE Transactions on Neural Networks*, *22*(6), 962–968.
19. Mangasarian, O. L. (1994). *Nonlinear programming*. Philadelphia: SIAM.
20. Chandra, S., Jayadeva, & Mehra, A. (2009). *Numerical optimization with applications*. New Delhi: Narosa Publishing House.
21. Saunders, C., Gammerman, A., & Vovk, V. (1998). Ridge regression learning algorithm in dual variables. *Proceedings of the Fifteenth International Conference on Machine Learning* (pp. 515–521).
22. Blake, C. L., & Merz, C. J. UCI Repository for Machine Learning Databases, Irvine, CA: University of California, Department of Information and Computer Sciences, http://www.ics.uci.edu/~mlearn/MLRepository.html.
23. Zheng, S. (2011). Gradient descent algorithms for quantile regression with smooth approximation. *International Journal of Machine Learning and Cybernetics*, *2*(3), 191–207.
24. Lagaris, I. E., Likas, A., & Fotiadis, D. (1998). Artificial neural networks for solving ordinary and partial differential equations. *IEEE Transactions on Neural Networks*, *9*, 987–1000.
25. Antonio, J., Martin, H., Santos, M., & Lope, J. (2010). Orthogonal variant moments features in image analysis. *Information Sciences*, *180*, 846–860.
26. Julier, S. J., & Uhlmann, J. K. (2004). Unscented filtering and nonlinear estimation. *Proceedings of the IEEE*, *92*, 401–422.
27. Lázarao, M., Santamaŕia, I., Péreze-Cruz, F., & Artés-Rodŕiguez, A. (2005). Support vector regression for the simultaneous learning of a multivariate function and its derivative. *Neurocomputing*, *69*, 42–61.
28. Lázarao, M., Santamaŕia, I., Péreze-Cruz, F., & Artés-Rodŕiguez, A. (2003). SVM for the simultaneous approximation of a function and its derivative. Proceedings of the IEEE international workshop on neural networks for signal processing (NNSP), Toulouse, France (pp. 189–198).
29. Jayadeva, Khemchandani, R., & Chandra, S. (2006). Regularized least squares twin SVR for the simultaneous learning of a function and its derivative, IJCNN, 1192–1197.
30. Jayadeva, Khemchandani, R., & Chandra, S. (2008). Regularized least squares support vector regression for the simultaneous learning of a function and its derivatives. *Information Sciences*, *178*, 3402–3414.
31. Khemchandani, R., Karpatne, A., & Chandra, S. (2013). Twin support vector regression for the simultaneous learning of a function and its derivatives. *International Journal of Machine Learning and Cybernetics*, *4*, 51–63.
32. Khemchandani, R., Karpatne, A., & Chandra, S. (2011). Generalized eigenvalue proximal support vector regressor. *Expert Systems with Applications*, *38*, 13136–13142.

33. Mangasarian, O. L., & Wild, E. W. (2006). Multisurface proximal support vector machine classification via generalized eigenvalues. *IEEE Transactions on Pattern Analysis and Machine Intelligence*, *28*(1), 69–74.
34. Tikhonov, A. N., & Arsenin, V. Y. (1977). *Solutions of Ill-posed problems*. New York: Wiley.
35. Parlett, B. N. (1998). *The symmetric eigenvalue problem: Classics in applied mathematics* (Vol. 20). Philadelphia: SIAM.
36. Guarracino, M. R., Cifarelli, C., Seref, O., & Pardalos, P. M. (2007). A classification method based on generalized eigenvalue problems. *Optimization Methods and Software*, *22*(1), 73–81.

Chapter 5
Variants of Twin Support Vector Machines: Some More Formulations

5.1 Introduction

In Chap. 3, we have presented the formulation of Twin Support Vector Machine (TWSVM) for binary data classification along with its two variants, namely the formulations of Twin Bounded Support Vector Machines (TBSVM) and Improved Twin Support Vector Machines (ITWSVM). These variants had addressed some of the possible computational issues related to TWSVM. Recently, many others variants/extensions of TWSVM have been proposed in the literature and we present some of these in this chapter. These are the Least Squares TWSVM (LS-TWSVM), ν-TWSVM, Parametric- TWSVM (par-TWSVM), Non-Parallel Plane classifier (NPPC), and many more. We would discuss pros and cons of these models and their usefulness in the TWSVM domain.

This chapter is organized as follows. Section 5.2 develops the details of Least Squares TWSVM and Sect. 5.3 discusses ν-Twin Support Vector Machine formulation. Section 5.4 deals with the parametric version of TWSVM while in Sect. 5.5 non-parallel version of TWSVM will be presented. Section 5.6 contains certain concluding remarks on these variants of TWSVM.

5.2 Least Squares-TWSVM

On the lines of Least Squares SVM proposed by Suykens and Vandewalle [1], Kumar and Gopal [2] presented Least Squares TWSVM (LS-TWSVM) where they modified the primal and dual QPPs of TWSVM in least squares sense and solved them with equality constraints instead of inequalities of TWSVM. These modifications simplified the solution methodology for LS-TWSVM as one could get its solution directly from solving two systems of linear equations instead of solving two QPPs as in TWSVM. Here we would restrict ourself to the linear version only, as the kernel version being analogous to kernel version of TWSVM.

Jayadeva et al., *Twin Support Vector Machines*, Studies in Computational Intelligence 659, DOI 10.1007/978-3-319-46186-1_5

Let the training set T_C for the given binary data classification problem be

$$T_C = \{(x^{(i)}, y_i), x^{(i)} \in \mathbb{R}^n, y_i \in \{-1, +1\}, (i = 1, 2, ..., m)\}. \tag{5.1}$$

Let there be m_1 patterns in class $+1$ and m_2 patterns in class -1 with $m_1 + m_2 = m$. We form the $(m_1 \times n)$ matrix A with its ith row as $(x^{(i)})^T$ for which $y_i = +1$. The $(m_2 \times n)$ matrix B is constructed similarly with data points $x^{(i)}$ for which $y_i = -1$.

The linear TWSVM formulation consists of determining two hyperplanes, namely positive hyperplane, $x^T w_1 + b_1 = 0$ and negative hyperplane, $x^T w_2 + b_2 = 0$ by solving the following pair of quadratic programming problems

$$(LS-TWSVM1) \quad \underset{(w_1,\, b_1,\, q_1)}{\text{Min}} \frac{1}{2}\|(Aw_1 + e_1 b_1)\|_2 + C_1 {q_1}^T q_1$$
$$\text{subject to}$$
$$-(Bw_1 + e_2 b_1) + q_1 = e_2, \tag{5.2}$$

and

$$(LS-TWSVM2) \quad \underset{(w_2,\, b_2,\, q_2)}{\text{Min}} \frac{1}{2}\|(Bw_2 + e_2 b_2)\|_2 + C_2 {q_2}^T q_2$$
$$\text{subject to}$$
$$(Aw_2 + e_1 b_2) + q_2 = e_1. \tag{5.3}$$

Here, $w_1 \in \mathbb{R}^n$, $w_2 \in \mathbb{R}^n$, $b_1 \in \mathbb{R}$, $b_2 \in \mathbb{R}$, $q_1 \in \mathbb{R}^{m_2}$ and $q_2 \in \mathbb{R}^{m_1}$. Also $C_1 > 0, C_2 > 0$ are parameters, and $e_1 \in \mathbb{R}^{m_1}$, $e_2 \in \mathbb{R}^{m_2}$ are vectors of 'ones', i.e. each component is 'one' only. Note that LS-TWSVM uses L_2-norm of slack variables instead of L_1-norm as in TWSVM which makes the constraints $q_1, q_2 \geq 0$ redundant.

Since (LS-TWSVM1) is a QPP with equality constraints only, its K.K.T conditions [3, 4] results in a system of linear equations. Further, the strict convexity of the objective function asserts that (LS-TWSVM1) has unique solution $(w_1, b_1)^T$. It can be verified that

$$[w_1 \;\; b_1]^T = -(F^T F + \frac{1}{C_1} E^T E)^{-1} F^T e_2, \tag{5.4}$$

with $E = [A \;\; e_1]$ and $F = [B \;\; e_2]$. Similarly for (LS-TWSVM2), we obtain its solution as

$$[w_2 \;\; b_2]^T = +(E^T E + \frac{1}{C_2} F^T F)^{-1} E^T e_1. \tag{5.5}$$

Thus, LS-TWSVM solves the classification problem using two matrix inverses, one for each hyperplane where $[w_1, b_1]^T$ and $[w_2, b_2]^T$ are determined as per Eqs. (5.4) and (5.5) respectively. Since the computational complexity of solving a system of m linear equations in n unknowns is $O(r^3)$ where r is the rank of $m \times n$ matrix and $r \leq \text{Min}(m, n)$, the LS-TWSVM is computationally faster than TWSVM. Gopal and Kumar [2] implemented LS-TWSVM on various real and synthetic datasets and compared its efficiency to that of TWSVM, GEPSVM and proximal SVM. For these details and other related issues we may refer to the their paper [2].

Recently, Nasiri et al. [5] proposed Energy-based model of least squares twin support vector machine (ELS-TWSVM) in activity recognition in which they changed the minimum unit distance constraint used in LS-TWSVM by some fixed energy parameter in order to reduce the effect of intrinsic noise of spatio-temporal features [6]. However, this energy parameter was explicitly chosen based on external observation of the user or could be part of the optimization problem.

The primal form of ELS-TWSVM formulation is given below

$$(ELS-TWSVM1) \quad \underset{(w_1,b_1,y_2)}{\text{Min}} \quad \frac{1}{2}||Aw_1+e_1b_1||^2+\frac{C_1}{2}y_2^Ty_2$$
subject to
$$-(Bw_1+e_2b_1)+y_2=E_1, \tag{5.6}$$

and

$$(ELS-TWSVM2) \quad \underset{(w_2,b_2,y_1)}{\text{Min}} \quad \frac{1}{2}||Bw_2+e_2b_2||^2+\frac{C_2}{2}y_1^Ty_1$$
subject to
$$(Aw_2+e_1b_2)+y_1=E_2, \tag{5.7}$$

where E_1 and E_2 are predefined energy parameters.

On the similar lines of LS-TWSVM, we obtain the solution of QPPs (5.6) and (5.7) as follows

$$[w_1 \;\; b_1]^T = -[c_1G^TG+H^TH]^{-1}[c_1G^TE_1], \tag{5.8}$$

and

$$[w_2 \;\; b_2]^T = [c_2H^TH+G^TG]^{-1}[c_2H^TE_2]. \tag{5.9}$$

A new data point $x \in R^n$ is assigned to class i ($i = +1$ or -1) using the following decision function,

$$f(x) = \begin{cases} +1, & \text{if } \dfrac{|x^Tw_1+e_1b_1|}{|x^Tw_2+e_2b_2|} \leq 1, \\ \\ -1, & \text{if } \dfrac{|x^Tw_1+e_1b_1|}{|x^Tw_2+e_2b_2|} \geq 1. \end{cases}$$

ELS-TWSVM takes advantage of prior knowledge available in human activity recognition problem about the uncertainty and intra-class variations and thus improves the performance of the activity recognition to some degree (Nasiri et al. [5]). However, the drawback is that energy parameter has to be externally fixed which sometimes leads to instability in problem formulations which effects the overall prediction accuracy of the system. We would next discuss the formulation where the energy parameter discussed above becomes part of the optimization problem.

5.3 Linear ν-TWSVM

In terms of generalization ability, the experimental results in Jayadeva et al. [7] have shown that the TWSVM compares favorably with the SVM and the GEPSVM. However, in contrast to the classical SVM, the TWSVM only aims at minimizing the empirical risk of training examples. For instance, to find the hyperplane which is a representative of class A (termed as positive hyperplane), minimizes simultaneously the L_2-norm empirical risk of samples of class A and the L_1-norm empirical risks of samples of class B with a trade-off parameter. Specifically, the corresponding QPP not only minimizes the distances from the hyperplane to the training samples of class A but also requires the distances from the other class to be at least one; otherwise, slack variables are added to avoid contradictions. This principle may lead to an increase in the number of support vectors (SVs) and reduces the generalization performance of the TWSVM.

In this section, we present ν-Twin Support Vector Machine (ν-TWSVM) formulation proposed by Peng [8]. It is motivated by ν-Support Vector Machine (ν-SVM) (Scholkopf et al. [9]) and has additional variables ρ_+ and ρ_- along with new parameters ν_1 and ν_2. Similar to ν-SVM, the parameters ν_1 and ν_2 in the ν-TWSVM control the bounds of fractions of the numbers of SVs and the margin errors.

The linear ν-TWSVM formulation consists of determining two hyperplanes, i.e. positive hyperplane $x^T w_1 + b_1 = \rho_+$ and negative hyperplane $x^T w_2 + b_2 = \rho_-$ by solving the following pair of quadratic programming problems

$$(\nu - TWSVM1) \quad \underset{(w_1,\, b_1,\, q_1, \rho_+)}{\text{Min}} \quad \frac{1}{2}\|(Aw_1 + e_1 b_1)\|_2 - \nu_1 \rho_+ + \frac{1}{m_1} e_2^T q_1$$

subject to

$$\begin{aligned} -(Bw_1 + e_2 b_1) + q_1 &\geq \rho_+, \\ \rho_+, q_1 &\geq 0, \end{aligned} \tag{5.10}$$

and

$$(\nu - TWSVM2) \quad \underset{(w_2,\, b_2,\, q_2, \rho_-)}{\text{Min}} \quad \frac{1}{2}\|(Bw_2 + e_2 b_2)\|_2 - \nu_2 \rho_- + \frac{1}{m_2} e_1^T q_2$$

subject to

$$\begin{aligned} (Aw_2 + e_2 b_2) + q_2 &\geq \rho_-, \\ \rho_-, q_2 &\geq 0. \end{aligned} \tag{5.11}$$

Here, $w_1 \in \mathbb{R}^n$, $w_2 \in \mathbb{R}^n$, $b_1 \in \mathbb{R}$, $b_2 \in \mathbb{R}$, $q_1 \in \mathbb{R}^{m_2}$ and $q_2 \in \mathbb{R}^{m_1}$. Also $\nu_1 > 0$, $\nu_2 > 0$ are parameters, and $e_1 \in \mathbb{R}^{m_1}$, $e_2 \in \mathbb{R}^{m_2}$ are vectors of 'ones', i.e. each component is 'one' only.

There are two additional variables ρ_+ and ρ_- that need to be optimized in (5.10) and (5.11) respectively. To understand the role of ρ_+ and ρ_-, we note that for all $q_{1_i} = 0$, $(i = 1, 2, \ldots, m_1)$ or $q_{2_j} = 0$, $(j \in 1, 2, \ldots, m_2)$, the samples from both the classes are separated by their representative hyperplane, with the margin $\dfrac{\rho_+}{\|{w_1}^T w_1\|}$ $\left(or \dfrac{\rho_-}{\|{w_2}^T w_2\|}\right)$. The adaptive ρ_+ and ρ_- effectively selects training examples which may become SVs, and hence would improve generalization performance as compared to TWSVM. Thus, one may expect that the ν-TWSVM obtains better generalization performance and fewer SVs than the TWSVM. We observe that ν-TWSVM is also approximately four times faster then the usual ν-SVM.

Now we consider $(\nu - TWSVM1)$ and use K.K.T conditions to get its dual as

$$
\begin{aligned}
(\nu - DTWSVM1)\ \underset{\alpha}{Max}\ \ & \frac{1}{2}\alpha^T G(H^T H)^{-1} G^T \alpha \\
\text{subject to}\ \ & \\
& e_2^T \alpha \geq \nu_1, \\
& 0 \leq \alpha \leq \frac{1}{m_2}.
\end{aligned} \tag{5.12}
$$

Here $H = [A \quad e_1]$, $G = [B \quad e_2]$. Once the Lagrange multiplier α is obtained, the decision vector $u = [w_1, b_1]^T$ is given by

$$
u = -(H^T H)^{-1} G^T \alpha. \tag{5.13}
$$

In a similar manner, we consider (ν-TWSVM2) and obtain its dual as

$$
\begin{aligned}
(\nu - DTWSVM2)\ \underset{\beta}{\text{Max}}\ \ & \frac{1}{2}\beta^T H(G^T G)^{-1} H^T \beta \\
\text{subject to}\ \ & \\
& e_2^T \beta \geq \nu_2, \\
& 0 \leq \beta \leq \frac{1}{m_1}.
\end{aligned} \tag{5.14}
$$

Again knowing the Lagrange multiplier β, the decision vector $v = [w_2, b_2]^T$ is given by

$$
v = (G^T G)^{-1} H^T \beta. \tag{5.15}
$$

The vectors u and v so determined give the separating hyperplanes

$$x^T w_1 + b_1 = 0 \quad \text{and} \quad x^T w_2 + b_2 = 0. \tag{5.16}$$

A new data point $x \in \mathbb{R}^n$ is assigned to class r $(r = 1, 2)$, depending on which of the two planes given by (5.16) it lies closer to. Thus

$$\text{class}(x) = \underset{(i=1,2)}{arg\ Min}(d_r(x)),$$

where,

$$d_r(x) = \frac{|x^T w_r + b_r|}{\|w_r\|}. \tag{5.17}$$

Remark 5.3.1 Similar to TWSVM, Peng in [8] observed that patterns of class -1 for which $0 < \alpha_i < C_1$, $(i = 1, 2 \ldots, m_2)$, lie on the hyperplane given by $x^T w_1 + b_1 = \rho_+$. We would further define such patterns of class -1 as support vectors of class $+1$ with respect to class -1, as they play a important role in determining the required plane. A similar definition of support vectors of class -1 with respect to class $+1$ follows analogously.

5.4 Linear Parametric-TWSVM

Generally, the classical SVM and its extensions assume that the noise level on training data is uniform throughout the domain, or at least, its functional dependency is known beforehand. The assumption of a uniform noise model, however, is not always satisfied. For instance, for the heteroscedastic noise structure, the amount of noise depends heteroscedastic noise on location. Recently, Hao [10] aimed at this shortcoming appearing in the classical SVM, and proposed a novel SVM model, called the Parametric-Margin ν-Support Vector Machine (par-ν-SVM), based on the ν-Support Vector Machine (ν-SVM). This par-ν-SVM finds a parametric-margin model of arbitrary shape. The parametric insensitive model is characterized by a learnable function $g(x)$, which is estimated by a new constrained optimization problem. This can be useful in many cases, especially when the data has heteroscedastic error structure, i.e., the noise strongly depends on the input value.

In this section, we present a twin parametric margin SVM (TPMSVM) proposed by Peng [11]. The proposed TPMSVM aims at generating two nonparallel hyperplanes such that each one determines the positive or negative parametric-margin hyperplane of the separating hyperplane. For this aim, similar to the TWSVM, the TPMSVM also solves two smaller sized QPPs instead of solving large one as in the classical SVM or par-ν-SVM. The formulation of TPMSVM is totally different from that of par-ν-SVM in some respects. First, the TPMSVM solves a pair of smaller sized QPPs, whereas, the par-ν-SVM only solves single large QPP, which makes the learning speed of TPMSVM much faster than the par-ν-SVM. Second, the par-ν-SVM directly finds the separating hyperplane and parametric-margin hyperplane, while the TPMSVM indirectly determines the separating hyperplane through the positive and negative parametric-margin hyperplanes. In short, TPMSVM successfully combines the merits of TWSVM, i.e., the fast learning speed, and par-ν-SVM, i.e., the flexible parametric-margin. In their paper, Peng [11] has done the computational comparisons of TPMSVM, par-ν-SVM, TWSVM and SVM in terms of generalization performance, number of support vectors (SVs) and training time are made on several artificial and benchmark datasets, indicating the TPMSVM is not only fast, but also shows comparable generalization.

To understand the approach of TPMSVM, we would first present par-ν-SVM model in order to explain the mathematics behind the parametric behaviour of SVM model. The par-ν-SVM considers a parametric-margin mode $g(x) = z^T x + d$ instead of the functional margin in the ν-SVM. Specifically, the hyperplane $f(x) = w^T x + b$ in the par-ν-SVM separates the data if and only if

$$x^T w + b \geq x^T z + d, \quad \forall\, x \in A, \tag{5.18}$$

$$x^T w + b \leq -x^T z - d, \quad \forall\, x \in B. \tag{5.19}$$

An intuitive geometric interpretation for the par-ν-SVM is shown in Fig. 5.1 (Hao [10]).

To find $f(x)$ and $g(x)$ along with the decision variables w and b, the par-ν-SVM considers the following constrained optimization problem

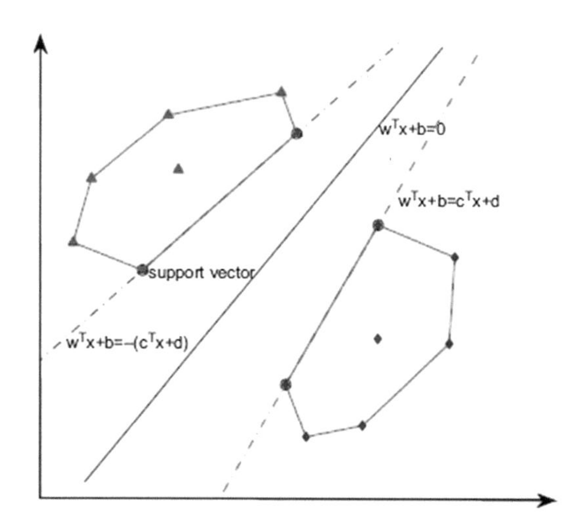

Fig. 5.1 Geometric interpretation for the par-ν-SVM

$$(par-SVM) \quad \underset{(w,\,b,\,q,\,z,\,d)}{\text{Min}} \quad \frac{1}{2}\|w\|_2^2 + c(\nu.(\frac{1}{2}\|z\|^2 + d) + \frac{1}{m}e^T q$$
subject to

$$y_i(w^T x^{(i)} + eb) \geq (z^T x^{(i)} + ed) - q, \qquad (i = 1, 2, \ldots, m),$$
$$d, q \geq 0. \tag{5.20}$$

Here, $w \in \mathbb{R}^n$, $z \in \mathbb{R}^n$, $b \in \mathbb{R}$, $d \in \mathbb{R}$, and $q \in \mathbb{R}$. Also $\nu > 0$ is a trade-off parameter, and $e \in \mathbb{R}^m$ are vectors of 'ones', i.e. each component is 'one' only.

By introducing the Lagrangian function, dual QPP for *par* $- \nu$-SVM is given by

$$(par-DSVM) \quad \underset{\alpha_i}{\text{Max}} \quad -\frac{1}{2}\sum_{i=1}^{m}\sum_{j=1}^{m} y_i y_j x^{(i)} x^{(j)} \alpha_i \alpha_j + \frac{1}{2c\nu}\sum_{i=1}^{m}\sum_{j=1}^{m} x^{(i)} x^{(j)} \alpha_i \alpha_j$$
subject to

$$e_2^T \alpha \geq c\nu,$$
$$\sum_{i=1}^{m} y_i \alpha_i = 0,$$
$$0 \leq \alpha_i \leq \frac{ce}{m}. \tag{5.21}$$

Solving the above dual QPP, we obtain the vector of Lagrange multipliers α, which gives the weight vectors w and z as the linear combinations of data samples

$$w = \sum_{i=1}^{m} y_i x^{(i)} \alpha_i, \qquad z = \frac{1}{c\nu}\sum_{i=1}^{m} x^{(i)} \alpha_i.$$

Further, by exploiting the K.K.T. conditions [3, 4] the bias terms b and d are determined as

$$b = -0.5 * [w^T x^{(i)} + w^T x^{(j)} - z^T x^{(i)} + w^T x^{(j)}],$$

and

$$d = 0.5 * [w^T x^{(i)} - w^T x^{(j)} - z^T x^{(i)} - w^T x^{(j)}].$$

Here, i and j are such that α_i corresponding to $y_i = 1$ and α_j corresponding to $y_j = -1$ satisfy $\alpha_i, \alpha_j \in (0, \frac{c}{m})$. Therefore, the separating hyperplane $f(x)$ and the parametric-margin function $g(x)$ of the resulting par-ν-SVM takes the form

$$f(x) = \sum_{i=1}^{m} \alpha_i y_i (x^{(i)})^T x + b,$$

and

$$g(x) = \frac{1}{cv} \sum_{i=1}^{m} \alpha_i (x^{(i)})^T x + d,$$

respectively.

In the following, we call the two nonparallel hyperplanes $f(x) \pm g(x) = 0$ as the parametric-margin hyperplanes.

Similar to the TWSVM, TPMSVM also derives a pair of nonparallel planes around the datapoints through two QPPs in order to find $f_1(x) = x^T w_1 + b_1 = 0$ and $f_2(x) = x^T w_2 + b_2 = 0$ each one determines one of parametric-margin hyperplanes.

Peng [11] in their paper addressed, $f_1(x)$ and $f_2(x)$ as the positive and negative parametric-margin hyperplanes, respectively. Specifically, given the training datapoints $\{(x^{(i)}, y_i), x^{(i)} \in \mathbb{R}^n, y_i \in \{-1, +1\}, (i = 1, 2, ..., m)\}$, $f_1(x)$ determines the positive parametric-margin hyperplane, and $f_2(x)$ determines the negative parametric-margin hyperplane. By incorporating the positive and negative parametric margin hyperplanes, this TPMSVM separates the data if and only if

$$x^T w_1 + b_1 \geq 0 \qquad \forall x \in A,$$

$$x^T w_2 + b_2 \leq 0 \qquad \forall x \in B.$$

It is further, discussed in the paper that the parametric margin hyperplanes $f_1(x) = 0$ and $f_2(x) = 0$ in TPMSVM are equivalent to $f(x) \pm g(x) = 0$ in the par-ν-SVM.

Thus, Peng [11] considered the following pair of constrained optimization problems:

$$(par-TWSVM1) \quad \underset{(w_1,\, b_1,\, q_1)}{\text{Min}} \quad \frac{1}{2}\|w_1\|_2^2 + \frac{\nu_1}{m_2} e_2^T (Bw_1 + e_2 b_1) + \frac{c_1}{m_1} e_2^T q_1$$

subject to,

$$\begin{aligned} Aw_1 + e_1 b_1 &\geq 0 - q_1, \\ q_1 &\geq 0, \end{aligned} \tag{5.22}$$

and

$$(par-TWSVM2) \quad \underset{(w_2,\, b_2,\, q_2)}{\text{Min}} \quad \frac{1}{2}\|w_2\|_2^2 - \frac{\nu_2}{m_1} e_1^T (Aw_2 + e_1 b_2) + \frac{c_2}{m_2} e_1^T q_2$$

subject to,

$$\begin{aligned} Bw_2 + e_2 b_2 &\geq 0 - q_2, \\ q_2 &\geq 0, \end{aligned} \tag{5.23}$$

where $c_1, c_2, \nu_1, \nu_2 > 0$ are the regularization parameters which determine the penalty weights, and q_1, q_2 are the slack vectors and m_1 and m_2 represent number of patterns in class A and B respectively.

Similar to the previous discussed approaches, introducing the Lagrangian and with the help of K.K.T conditions, would lead to the following dual problems corresponding to their primal versions.

$$(par-DTWSVM1) \ \underset{\alpha}{\text{Max}} \quad -\frac{1}{2}\alpha^T AA^T\alpha + \frac{\nu_1}{m_1}e_2^T BA^T\alpha$$
subject to,
$$e_2^T\alpha = \nu_1,$$
$$0 \le \alpha \le \frac{c_1}{m_2}e_2. \tag{5.24}$$

Once the vector of dual variables α is obtained, the vector w_1 and b_1 are determined by $w_1 = A^T\alpha - \frac{\nu_1}{m_2}Be_2$ and $b_1 = -\frac{1}{|\mathbf{N}_+|}\sum_{i\in\mathbf{N}_+} A_i w_2$ where N_+ is the index set of positive samples satisfying $\alpha_i \in (0, \frac{c_1}{m_1})$ for some i.

Similarly we obtain the dual QPP of (par-TWSVM2) as

$$(par-DTWSVM2) \ \underset{\alpha}{\text{Max}} \quad -\frac{1}{2}\beta^T BB^T\beta + \frac{\nu_2}{m_2}e_1^T AA^T\beta$$
subject to
$$e_2^T\beta = \nu_2,$$
$$0 \le \beta \le \frac{c_2}{m_2}e_2. \tag{5.25}$$

On the similar lines, the vector w_2 and b_2 are determined by

$$w_2 = -B^T\beta + \frac{\nu_2}{m_1}A^T e_1,$$

and

$$b_2 = -\frac{1}{|\mathbf{N}_-|}\sum_{j\in\mathbf{N}_-} B_j w_2,$$

where N_- is the index set of positive samples satisfying $\beta_j \in (0, \frac{c_2}{m_2})$.

On the similar lines, using the concept of structural granularity, Peng et al. [12] introduced improved version of TPMSVM, termed as Structural TPMSVM. Structural TPMSVM incorporates data structural information within the corresponding class by adding regularization term derived from cluster granularity.

On the similar lines of Hao et al. [10], Khemchandani and Sharma [13] proposed Robust parametric TWSVM and have shown its application in human activity recognition framework.

Remark 5.4.1 The TWSVM classifier finds, for each class, a hyperplane that passes through the points of that class and is at a distance of at least unity from the other class. In contrast to the TWSVM classifier, par-TWSVM finds a plane that touches the points of one class so that points of that class lie on the one side, and is as far away as possible from points of the other one. Thus here the role is reversed and therefore par-TWSVM is more in the spirit of reverse twin rather than twin spirit.

5.5 Non-parallel Support Vector Machines

In Chap. 3, we have noticed that for the nonlinear case, TWSVM considers the kernel generated surfaces instead of hyperplanes and construct two different primal problems, which means that TWSVM has to solve two problems for linear case and two other problems for the nonlinear case separately. However, in the standard SVMs, only one dual problem is solved for both the cases with different kernels.

In order to address this issue, Tian et al. [14] proposed a novel nonparallel SVM, termed as NPSVM for binary classification where the dual problems of these two primal problems have the same advantages as that of the standard SVMs, i.e., only the inner products appear so that the kernel trick can be applied directly. Further, the dual problems have the same formulation with that of standard SVMs and can certainly be solved efficiently by SMO, i.e. no need to compute the inverses of the large matrices as we do in TWSVMs so as to solve the dual problem.

On the similar lines of TWSVM, NPSVM also seek two nonparallel hyperplanes by solving two convex QPPs.

$$(NPSVM1) \quad \underset{(w_1,\, b_1,\, \eta_1,\, \eta_1^*,\, q_1)}{\text{Min}} \quad \frac{1}{2}\|w_1\|^2 + C_1(e_1^T\eta_1 + e_1^T\eta_1^*) + C_2(e_2^Tq_1)$$

subject to

$$\begin{aligned} Aw_1 + e_1b_1 &\leq \epsilon + \eta_1, \\ -(Aw_1 + e_1b_1) &\leq \epsilon + \eta_1^*, \\ Bw_1 + e_2b_1 &\leq -e_2 + q_1, \\ \eta_1, \eta_1^*, q_1 &\geq 0, \end{aligned} \tag{5.26}$$

and

$$(NPSVM2) \quad \underset{(w_2,\, b_2,\, \eta_2,\, \eta_2^*,\, q_2)}{\text{Min}} \quad \frac{1}{2}\|w_2\|^2 + C_3(e_2^T\eta_2 + e_2^T\eta_2^*) + C_4(e_1^Tq_2)$$

subject to

$$\begin{aligned} Bw_2 + e_2b_2 &\le \epsilon + \eta_2, \\ -(Bw_2 + e_2b_2) &\le \epsilon + \eta_2^*, \\ Aw_2 + e_1b_2 &\ge -e_1 + q_2, \\ \eta_2, \eta_2^*, q_2 &\ge 0, \end{aligned} \tag{5.27}$$

where $C_1, C_2, C_3, C_4 > 0$ are the regularization parameters which determine the penalty weights, and $q_1, q_2, \eta_1, \eta_2, \eta_1^*, \eta_2^*$ are the slack vectors.

We first illustrate the physical interpretation of first problem where the first terms tends positive class to locate as much as possible in the ϵ-band between the hyperplanes $x^T w_1 + b_1 = \epsilon$ and $x^T w_2 + b_2 = -\epsilon$, the errors η_1 and η_1^* measures the ϵ-insensitive loss function and the second term maximizes the margin between the hyperplanes $x^T w_1 + b_1 = \epsilon$ and $x^T w_1 + b_1 = -\epsilon$, which can be expressed by $\frac{2\epsilon}{\|w\|}$. Further, similar with the TWSVM, the constraints pushes the negative class from the hyperplane $x^T w_1 + b_1 = -1$ as far as possible, the corresponding errors are measured by the soft margin loss function.

The corresponding dual problem are as follows

$$(DNPSVM1)\ \underset{(\alpha,\alpha^*,\beta)}{\text{Max}} \quad -\frac{1}{2}(\alpha^* - \alpha)^T AA^T(\alpha^* - \alpha) + (\alpha^* - \alpha)\beta BA^T$$
$$-\frac{1}{2}\beta^T BB^T\beta + \epsilon e_1^T(\alpha^* - \alpha) - e_2^T\beta$$

subject to

$$\begin{aligned} e_1^T(\alpha - \alpha^*) + e_2^T\beta &= 0, \\ 0 \le \alpha, \alpha^* &\le c_1, \\ 0 \le \beta &\le c_2, \end{aligned} \tag{5.28}$$

and

$$(DNPSVM2)\ \underset{(\alpha,\alpha^*,\beta)}{\text{Max}} \quad -\frac{1}{2}(\alpha^* - \alpha)^T BB^T(\alpha^* - \alpha) + (\alpha^* - \alpha)\beta AB^T$$
$$-\frac{1}{2}\beta^T AA^T\beta + \epsilon e_2^T(\alpha^* - \alpha) - e_1^T\beta$$

subject to

$$\begin{aligned} e_2^T(\alpha - \alpha^*) - e_1^T\beta &= 0, \\ 0 \le \alpha, \alpha^* &\le c_3, \\ 0 \le \beta &\le c_4, \end{aligned} \tag{5.29}$$

where α, α^* and β are the vector of Lagrange multipliers.

Once the solution in terms of w and b is obtained the decision rule is similar to that of TWSVM. Further, NPSVM degenerates to TBSVM and TWSVM when parameters are chosen appropriately. Interested readers could refer to Tian et al. [14] for more details.

5.6 Multi-category Extensions of TWSVM

SVM has been widely studied as binary classifiers and researcher have been trying to extend the same to multi-category classification problems. The two most popular approaches for multi-class SVM are One-Against-All (OAA) and One-Against-One (OAO) SVM (Hsu and Lin [15]). OAA-SVM implements a series of binary classifiers where each classifier separates one class from rest of the classes. But this approach leads to biased classification due to huge difference in the number of samples. For a K-class classification problem, OAA-SVM requires K binary SVM comparisons for each test data. In case of OAO-SVM, the binary SVM classifiers are determined using a pair of classes at a time. So, it formulates upto $(K * (K - 1))/2$ binary SVM classifiers, thus leading to increase in computational complexity. Also, directed acyclic graph SVM (DAGSVM) is proposed in Platt et al. [16], in which the training phase is the same as OAO-SVM i.e. solving $(K * (K - 1))/2$ binary SVM, however its testing phase is different. During testing phase, it uses a rooted binary directed acyclic graph which has $(K * (K - 1))/2$ internal nodes and K leaves. OAA-SVM classification using decision tree was proposed by Kumar and Gopal in ([17]). Chen et al. proposed multiclass support vector classification via coding and regression (Chen et al. [18]). Jayadeva et al. proposed fuzzy linear proximal SVM for multi-category data classification (Jayadeva et al. [19]). Lei et al. propose Half-Against-Half (HAH) multiclass-SVM (Lie and Govindarajen [20]). HAH is built via recursively dividing the training dataset of K classes into two subsets of classes. Shao et al. propose a decision tree twin support vector machine (DTTSVM) for multi-class classification (Shao et al. [21]), by constructing a binary based on the best separating principle, which maximizes the distance between the classes. Xie et al. have extended TWSVM for multi-class classification (Xie et al. [22]) using OAA approach. Xu et al. proposed Twin K-class support vector classifier (TwinKSVC) (Xu et al. [23]), which uses TSVM with support vector classification-regression machine for K-class classification (K-SVCR) and evaluates all the training points into a 1-versus-1-versus structure thereby generating ternary outputs $(+1, 0, -1)$.

The speed while learning a model is a major challenge for multi-class classification problems in SVM. Also, TWSVM classifier is four times faster than that of SVM, while learning a model, as it solves two smaller QPPs. Further, TWSVM overcomes the unbalance problem in two classes sized by choosing two different penalty variables for different classes. Because of the strength of TWSVM, Khemchandani and Saigal [24] have recently extended TWSVMs to multi-category scenario and termed it as Ternary Decision Structure based Multi-category Twin Support Vector

Machine (TDS-TWSVM) classifier. The key features of TDS-TWSVM are listed below:

- TDS-TWSVM determines a decision structure of TWSVM classifiers using training data. Each decision node is split into three decision nodes labeled as $(+1, 0, -1)$, where $+1$, and -1 represent focused groups of classes and 0 represents ambiguous group of classes. Ambiguous group consists of training samples with low confidence.
- At each level of the decision structure, a K-class problem is partitioned into atmost three $K/3$-class problems, until all samples belong to only one class. Thus it evaluates all the training samples into i-versus-j-versus-rest structure, where $i, j < K$.
- TDS-TWSVM requires $\lceil log_3 K \rceil$ tests, on an average, for evaluation of test sample.

During the training phase, TDS-TWSVM recursively divides the training data into three groups by applying k-means (k = 2) clustering [25] and creates a ternary decision structure of TWSVM classifiers, as shown in Fig. 5.2. The training set is first partitioned into two clusters which leads to identification of two focused groups of classes and an ambiguous group of classes. The focused class is one where most of the samples belong to a single cluster whereas the samples of an ambiguous group are scattered in both the clusters. Therefore, proposed algorithm in [24] has ternary outputs $(+1, 0, -1)$. TDS-TWSVM partitions each node of the decision structure into at most three groups, as shown in Fig. 5.3. The cluster labels $+1$, 0, -1 are assigned to training data and three hyperplanes are determined using one-against-all approach. This in turn creates a decision structure with height $\lceil log_3 K \rceil$. Thus, TDS-TWSVM is an improvement over OAA multiclass TWSVM approach with respect to training time of classifier and retrieval accuracy. The class of test data is evaluated by assigning the label of the class corresponding to which test data has minimum distance. The hyperplanes, thus obtained, are represented by nodes of the ternary decision structure and the K non-divisible nodes represent K-classes. This dynamic arrangement of classifiers significantly reduces the number of tests required in testing phase. With

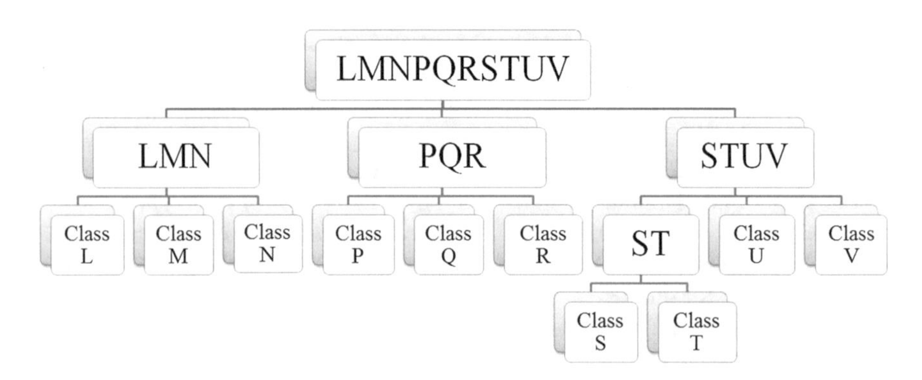

Fig. 5.2 Ternary decision structure of classifiers with 10 classes

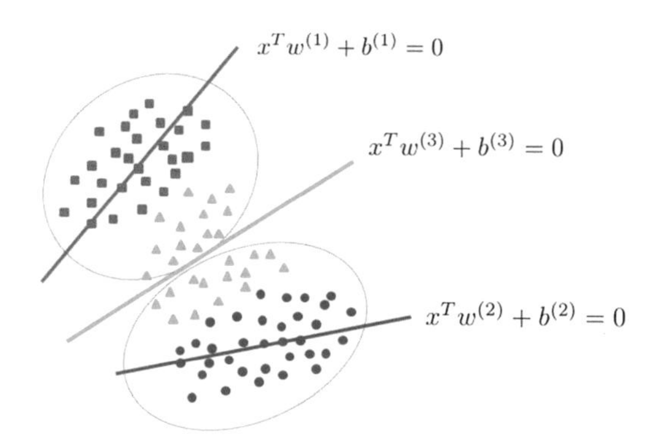

Fig. 5.3 Illustration of TDS-TWSVM

a balanced ternary structure, a K-class problem would require only $\lceil log_3 K \rceil$ tests. Also, at each level, the number of samples used by TDS-TWSVM diminishes with the expansion of decision structure. Hence, the order of QPP reduces as the height of the structure increases. The proposed TDS-TWSVM algorithm determines the classifier model, which is efficient in terms of accuracy and requires fewer tests for a K-class classification problem. The process of finding TDS-TWSVM classifier is explained below in the algorithm.

Algorithm: TDS-TWSVM

(This structure can be applied in general to any type of dataset; however experiments are performed in context of image classification). Given an image dataset with N images from K different classes. Pre-compute the Complete Binary Local Binary Pattern with Co-nonoccurrence matrix (CR-LBP-Co) [26] and Angular Radial Transform (ART) [27] features for all images in the dataset as discussed in Sects. 3.1 and 3.2 respectively. Create a descriptor F by concatenating both the features. F is a matrix of size $N \times n$, where n is the length of feature vector. Here, $n = 172$ and the feature vector for an image is given as
$fv = [ft_1, ft_2, ..., ft_{136}, fs_1, fs_2, ..., fs_{36}]$,
where $ft_i (i = 1, 2, ..., 136)$ is texture feature and $fs_k (k = 1, 2, ..., 36)$ is shape feature.

1. Select the parameters such as- penalty parameter C_i, kernel type and kernel parameter.
   Repeat Steps 2-5, 5-times (for 5-fold crossvalidation)
2. Use k-means clustering to partition the training data into two sets. Identify two focused groups of classes with labels '+1' and '−1' respectively, and one ambiguous group of classes represented with label '0'. Here, k=2 and we get at most three groups.

3. Take training samples of ‘+1’, ‘−1’ and ‘0’ groups as class representatives and find three hyperplanes $(w^{(1)}, b^{(1)})$, $(w^{(2)}, b^{(2)})$ and $(w^{(3)}, b^{(3)})$, by applying one-against-all approach and solving for TWSVM classifier.
4. Recursively partition the data-sets and obtain TWSVM classifiers until further partitioning is not possible.
5. Evaluate the test samples with the decision structure based classifier model and assign the label of the non-divisible node.

5.6.1 Other Multi-category Twin Support Vector Machine Approaches

The strength of the proposed algorithm lies in the fact that it requires fewer number of TWSVM comparisons for evaluations than other state-of-the-art multi-class approaches like OAA-SVM and OAO-SVM. In order to compare the accuracy of the proposed system, Khemchandani and Saigal [24] have implemented OAA-TWSVM and TB-TWSVM. OAA-TWSVM consists of solving K QPPs, one for each class, so that we obtain $2 * K$ nonparallel hyperplanes for K classes. Here, we construct a TWSVM classifier, where in i^{th} TWSVM classifier, we solve one QPP taking i^{th} class samples as one class and remaining samples as other class. By using the TWSVM methodology, we determine the hyperplane for the ith class. The unbalance problem of exemplars existing in i^{th} TWSVM is tackled by choosing the proper penalty variable C_i for the ith class. In case of TB-TWSVM, we recursively divide the data into two halves and create a binary tree of TWSVM classifiers. TB-TWSVM determines $2 * (K - 1)$ TWSVM classifiers for a K-class problem. For testing, TB-TWSVM requires at most $\lceil log_2 K \rceil$ binary TWSVM evaluations. They also implemented a variation of TB-TWSVM as ternary tree-based TWSVM (TT-TWSVM) where each node of the tree is recursively divided into three nodes. The partitioning is done by k-means clustering, with k=3. The experimental results show that TDS-TWSVM outperforms TT-TWSVM.

TDS-TWSVM is more efficient than OAA-TWSVM considering the time required to build the multi-category classifier. Also new test sample can be tested by $\lceil log_3 K \rceil$ comparisons, which is more efficient than OAA-TWSVM and TB-TWSVM testing time. For a balanced decision structure, the order of QPP reduces to one-third of parent QPP with each level, as the classes of parent node are divided into three groups. Experimental results show that TDS-TWSVM has advantages over OAA-TWSVM, TT-TWSVM and TB-TWSVM in terms of testing complexity. At the same time, TDS-TWSVM outperforms other approaches in multi-category image classification and retrieval.

5.6.2 *Content-Based Image Classification Using TDS-TWSVM*

Multicategory image classification is an automated technique of associating a class label with an image, based on its visual content. The multi-category classification task includes training a classifier model for all the image classes and evaluating the performance of the classifier by computing its accuracy on unseen (out-of-sample) data. Classification includes a broad range of decision-based approaches for identification of images. These algorithms are based on the assumption that the images possess one or more features and these features are associated to one of several distinct and exclusive classes.

For TDS-TWSVM based image classification, Khemchandani and Saigal [24] divided the image dataset into training and test data. To avoid overfitting, they have used 5-fold cross-validation. They randomly partitioned the dataset into five equal-sized sub-samples. Of these five sub-samples, one is retained as the evaluation set for testing the model, and the remaining four sub-samples are used as training data. The proposed algorithm works on image datasets with multiple classes. The training data is used to determine a classifier model and each test sample is evaluated using this model based on minimum euclidean distance from the three hyperplanes at each level of the classifier structure, until it reaches a non-divisible node. The label of leaf node is assigned to this test sample. The accuracy of the model is calculated by taking the average accuracy over all the folds with standard deviation. An important application of image classification is image retrieval i.e. searching through an image dataset to retrieve best matches for a given query image, using their visual content.

5.6.3 *Content-Based Image Retrieval Using TDS-TWSVM*

CBIR makes use of image features to determine the similarity or distance between two images [28]. For retrieval, CBIR fetches most similar images to the given query image. Khemchandani and Saigal [24] proposed the use of TDS-TWSVM for image retrieval. To find the class label of query image, the algorithm is explained in Sect. 5.6 and then to find the similar images from the classified training set, chi-square distance measure is used. A highlighting feature of TDS-TWSVM is that it is evaluated using out-of-sample data. Most CBIR approaches take query image from the dataset which is used to determine the model. But unlike other CBIR based approaches, TDS-TWSVM reserves a separate part of dataset for evaluation. Thus, it provides a way to test the model on data that has not been a component in the optimization model. Therefore, the classifier model will not be influenced in any way by the out-of-sample data. Figure 5.4 shows the retrieval result for a query image taken from Wang's Color dataset.

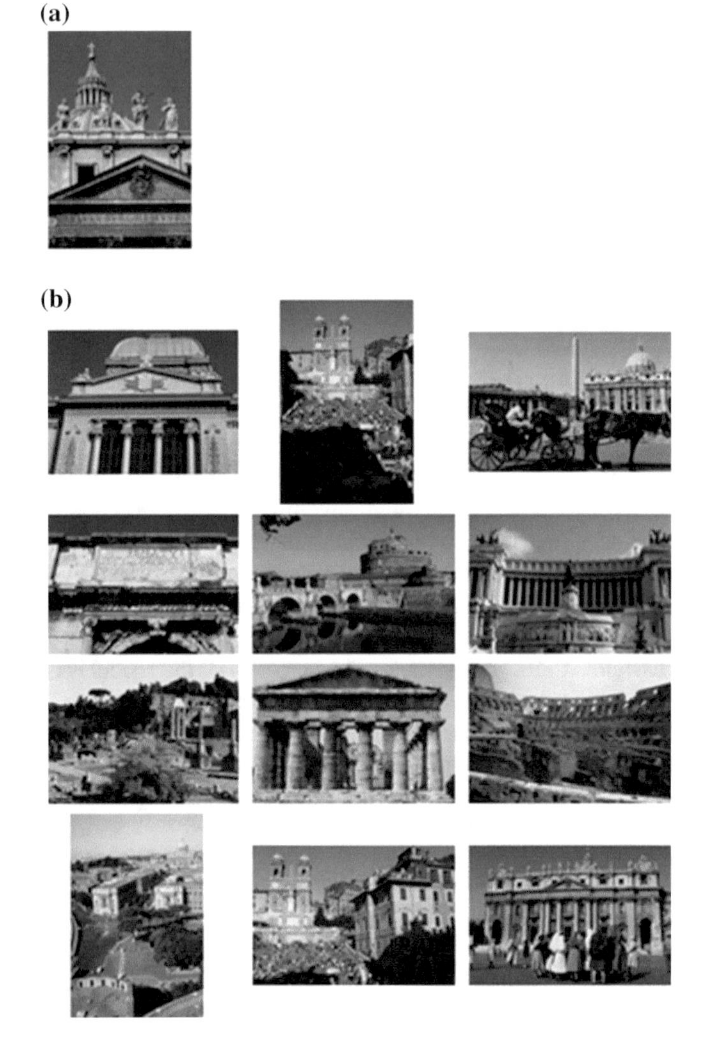

Fig. 5.4 Retrieval result for a query image taken from Wang's Color dataset

5.7 Conclusions

In this chapter, we review the variants of TWSVM which are developed in the recent past. The chapter discusses least squares version of TWSVM which is faster than TWSVM as it solves system of linear equations for obtaining the solution of TWSVM.

Further, ν-TWSVM was discussed where the unit distance separation constraint in TWSVM is relaxed to the parameter ρ which becomes part of optimization problem.

To capture heteroscedastic noise and outlier, par-TWSVM was also reviewed in this chapter. In this approach learnable function automatically adjusts a flexible parametric-insensitive zone of arbitrary shape and minimal radius to include the given data.

Towards the end, nonparallel classifier termed as NPSVM was introduced where ϵ-insensitive loss function instead of quadratic loss function is used. NPSVM implements the SRM principle, further the dual problem have same advantage as that of standard SVMs and can be solved efficiently by the SMO algorithm in order to deal with large scale problems.

The other variants of TWSVM which we have not discussed in this chapter are Fuzzy Twin Support Vector Machines (Khemchandani et al. [29]), Incremental Twin Support Vector Machines (Khemchandani et al. [30]), Probabilistic TWSVM(Shao et al. [31]), ν-Nonparallel Support Vector Machines (Tian et al. [32]), efficient sparse nonparallel support vector machines (Tian et al. [33]) and Recursive Projection TWSVM ([34]) etc. Interested readers could refer to some more related papers to aforementioned papers. In this regard certain survey articles on TWSVM for example Tian et al. [35], Ding et al. [36, 37] are good source of information.

References

1. Suykens, J. A., & Vandewalle, J. (1999). Least squares support vector machine classifiers. *Neural processing letters*, *9*(3), 293–300.
2. Kumar, M. A., & Gopal, M. (2009). Least squares twin support vector machines for pattern classification. *Expert Systems and its Applications*, *36*(4), 7535–7543.
3. Chandra, S., Jayadeva, & Mehra, A. (2009). *Numerical Optimization with Applications*. New Delhi: Narosa Publishing House.
4. Mangasarian, O. L. (1994). *Nonlinear Programming*. Philadelphia: SIAM.
5. Nasiri, J. A., Charkari, N. M., & Mozafari, K. (2014). Energy-based model of least squares twin Support Vector Machines for human action recognition. *Signal Processing*, *104*, 248–257.
6. Laptev I., Marszalek M., Schmid C., & Rozenfeld B. (2008). Learning realistic human actions from movies. *IEEE Conference on Computer Vision and Pattern Recognition, CVPR*, (p. 18). IEEE.
7. Jayadeva, Khemchandani. R., & Chandra, S. (2007). Twin support vector machines for pattern classification. *IEEE Transactions on Pattern Analysis and Machine Intelligence*, *29*(5), 905–910.
8. Peng, X. J. (2010). A ν-twin support vector machine (ν-TWSVM) classifier and its geometric algorithms. *Information Science*, *180*(20), 3863–3875.
9. Schoolkopf, B., Smola, A., Williamson, R., & Bartlett, P. L. (2000). *New support vector algorithms. Neural Computation*, *12*(5), 1207–1245.
10. Hao, Y. P. (2010). New support vector algorithms with parametric insensitive margin model. *Neural Networks*, *23*(1), 60–73.
11. Peng, X. (2011). TPSVM: A novel twin parametric-margin support vector for pattern recognition. *Pattern Recognition*, *44*(10–11), 2678–2692.
12. Peng, X. J., Wang, Y. F., & Xu, D. (2013). Structural twin parametric margin support vector machine for binary classification. *Knowledge-Based Systems*, *49*, 63–72.

13. Khemchandani, R., & Sharma,S. (2016). Robust parametric twin support vector machines and its applications to human activity recognition. In *Proceedings of International Conference on Image Processing*, IIT Roorkee.
14. Tian, Y. J., Qi, Z. Q., Ju, X. C., Shi, Y., & Liu, X. H. (2013). Nonparallel support vector machines for pattern classification. *IEEE Transactions on cybernertics*, *44*(7), 1067–1079.
15. Hsu, C.-W., & Lin, C.-J. (2002). A comparison of methods for multiclass support vector machines. *IEEE Transactions on Neural Networks*, *13*(2), 415–425.
16. Platt, J. C., Cristianini, N., & Shawe-Taylor, J. (2010). Large margin DAGs for multiclass classification. *Advances in Neural Information Processing Systems*, *12*, 547–553.
17. Kumar, M. A., & Gopal, M. (2010). Fast multiclass SVM classification using decision tree based one-against-all method. *Neural Processing Letters*, *32*, 311–323.
18. Chen, P.-C., Lee, K.-Y., Lee, T.-J., Lee, Y.-J., & Huang, S.-Y. (2010). Multiclass support vector classification via coding and regression. *Neurocomputing*, *73*, 1501–1512.
19. Jayadeva, Khemchandani. R., & Chandra, S. (2005). Fuzzy linear proximal support vector machines for multi-category data classification. *Neurocomputing*, *67*, 426–435.
20. Lei, H., & Govindaraju, V. (2005). Half-against-half multi-class support vector machines. *MCS, LNCS*, *3541*, 156–164.
21. Shao, Y.-H., Chen, W.-J., Huang, W.-B., Yang, Z.-M., & Deng, N.-Y. (2013). The best separating decision tree twin support vector machine for multi-class classification. *Procedia Computer Science*, *17*, 1032–1038.
22. Xie, J., Hone, K., Xie, W., Gao, X., Shi, Y., & Liu, X. (2013). Extending twin support vector machine classifier for multi-category classification problems. *Intelligent Data Analysis*, *17*, 649–664.
23. Xu, Y., Guo, R., & Wang, L. (2013). A Twin multi-class classification support vector machine. *Cognate Computer*, *5*, 580–588.
24. Khemchandani, R., & Saigal, P. (2015). Color image classification and retrieval through ternary decision structure based multi-category TWSVM. *Neurocomputing*, *165*, 444–455.
25. Queen, M. J. (1967). Some methods for classification and analysis of multivariate observations. In, *Proceedings of 5th Berkeley Symposium on Mathematical Statistics and Probability* (pp. 281–297). University of California.
26. Zhao, Y., Jia, W., Hu, R. X., & Min, H. (2013). Completed robust local binary pattern for texture classification. *Neurocomputing*, *106*, 68–76.
27. Ricard, Julien, Coeurjolly, David, & Baskurt, Atilla. (2005). Generalizations of angular radial transform for 2D and 3D shape retrieval. *Pattern Recognition Letters*, *26*(14), 2174–2186.
28. Liu, G. H., Zhang, L., Hou, Y. K., Li, Z. Y., & Yang, J. Y. (2010). Image retrieval based on multi-texton histogram. *Pattern Recognition*, *43*(7), 2380–2389.
29. Khemchandani, R., Jayadeva, & Chandra, S. (2007). Fuzzy twin support vector machines for pattern classification. In *ISPDM' 07 International Symposium on Mathematical Programming for Decision Making: Theory and Applications*. Singapore: World Scientific (Published in Mathematical Programming and Game Theory for Decision Making)
30. Khemchandani, R., Jayadeva, & Chandra, S. (2008). Incremental twin support vector machines. In S.K. Neogy, A.K.das and R. B. Bapat (Eds.), *ICMCO-08, International Conference on Modeling, Computation and Optimization*. Published in Modeling, Computation and Optimization. Singapore:World Scientific.
31. Shao, Y. H., Deng, N. Y., Yang, Z. M., Chen, W. J., & Wang, Z. (2012). Probabilistic outputs for twin support vector machines. *Knowledge-Based Systems*, *33*, 145–151.
32. Tian, Y. J., Zhang, Q., & Liu, D. L. (2014). ν-Nonparallel support vector machine for pattern classification. *Neural Computing and Applications*,. doi:10.1007/s00521-014-1575-3.
33. Tian, Y. J., Ju, X. C., & Qi, Z. Q. (2013). Efficient sparse nonparallel support vector machines for classification. *Neural Computing and Applications*, *24*(5), 1089–1099.
34. Chen, X., Yang, J., Ye, Q., & Liang, J. (2011). Recursive projection twin support vector machine via within-class variance Minimization. *Pattern Recognition*, *44*(10), 2643–2655.
35. Tian, Y., & Qi, Z. (2014). Review on twin support vector machines Arin: Data. *Sci*,. doi:10.1007/S40745-014-0018-4.

36. Ding, S., Yu, J., Qi, B., & Huang, H. (2014). An Overview of twin support vector machines. *Artificial Intelligence Review.*, *42*(2), 245–252.
37. Ding, S., Zhang, N., Zhang, X., & Wu. F. (2016). Twin support vector machine: theory, algorithm and applications. *Neural Computing and Applications*,. doi:10.1007/s00521-016-2245-4.

Chapter 6
TWSVM for Unsupervised and Semi-supervised Learning

6.1 Introduction

Recently, Semi-Supervised Learning (SSL) has attracted interest of many researchers due to the availability of large amount of unlabeled examples as compared to rare examples with labels. In order to make any use of unlabeled data, we must assume some structure to the underlying distribution of data. Three main assumption which semi-supervised algorithms possess are

1. Smoothness: Points which are close to each other are more likely to share same label. This is also generally assumed in supervised learning and yields a preference for geometrically simple decision boundaries. In the case of semi-supervised learning, the smoothness assumption additionally yields a preference for decision boundaries in low-density regions, where there are fewer points close to each other but are from different classes.
2. Cluster: The data tend to form discrete clusters, and points in the same cluster are more likely to share common label (although data sharing a label may be spread across multiple clusters). This is a special case of the smoothness assumption and gives rise to feature learning with clustering algorithms.
3. Manifold: The data lie approximately on a manifold of much lower dimension than the input space. In this case attempts are made to learn the manifold using both the labeled and unlabeled data so as to avoid the curse of dimensionality. Then learning can proceed using distances and densities defined on the manifold.

Several novel approaches for utilizing the information of unlabeled data while building the classifier have been proposed by Belkin et al. [1], Blum and Mitchell [2], Nigam et al. [3]. Using graph Laplacian, Melacci et al. in [4] have proposed Laplacian Support Vector Machine (Lap-SVM) for semi-supervised learning. The main assumptions of these models are that all points are located in a low dimensional manifold and the graph is used for an approximation of the underlying manifold (Zhu [5]).

Jayadeva et al., *Twin Support Vector Machines*, Studies in Computational Intelligence 659, DOI 10.1007/978-3-319-46186-1_6

Recently, Jayadeva et al. [6] have proposed Twin Support Vector Machine (TWSVM) classifier for binary data classification where the two hyperplanes are obtained by solving two related smaller sized Quadratic Programming Problems (QPPs) as compare to single large sized QPP in conventional Support Vector Machine (SVM). Taking motivation from Suykens and Vanderwalla [7], Kumar and Gopal [8] have proposed Least Squares version of TWSVM (LS-TWSVM) and shown that the formulation is extremely fast since it solves two modified primal problem of TWSVM which is further equivalent to solving system of linear equations. Most researchers minimize the loss function subject to the L_1-norm and L_2-norm penalty. In [9] the authors proposed L_p-norm Least Square Twin Support Vector Machine (L_p-LSTWSVM) which automatically select the value of p from the data.

Extension of TWSVM in semi-supervised framework has been proposed by Qi et al. [10] which they termed as Laplacian-TWSVM. In Laplacian-TWSVM, the authors have used graph based method for utilizing labeled information along with large number of unlabeled information to build a better classifier. To reduce the computation cost of Laplacian-TWSVM, Chen et al. [11] proposed a least squares version of Laplacian-TWSVM, termed as Lap-LSTWSVM. Lap-LSTWSVM replaces the QPPs in Lap-TWSVM with a linear system of equations by using a squared loss function instead of the hinge loss function. Similar to LS-TWSVM, Lap-LSTWSVM is extremely fast as their solution is determined by solving system of linear equations.

In semi-supervised binary classification problem, we consider set $S = \{(x_1, y_1), (x_2, y_2), \ldots, (x_l, y_l), x_{l+1}, \ldots, x_m\}$, where $X_l = \{x_i : i = 1, 2, \ldots, l\}$ are the l labeled data points in n dimension with corresponding class labels $Y_l = \{y_i \in [1, -1] : i = 1, 2, \ldots, l\}$ and $X_u = \{x_i : i = l+1, l+2, \ldots, m\}$ are unlabeled data points. Thus, $X = X_l \cup X_u$. Data points belonging to class 1 and -1 are represented by matrices A and B each with number of patterns m_1 and m_2, respectively. Therefore, the size of matrices A and B are $(m_1 \times d)$ and $(m_2 \times d)$, respectively. Here, n is the dimension of the feature space. Let A_i $(i = 1, 2, \ldots, m_1)$ is a row vector in n-dimensional real space R^n that represents feature vector of data sample.

For better generalization performance, we would like to construct a classifier which utilizes both labeled and unlabeled data information. Recently, Manifold Regularization learning technique have been proposed by Belkin et al. [1] which utilizes both labeled and unlabeled data information as well as preserve some geometric information of data. In [1], the author introduced the following regularization term

$$||f||_M^2 = \frac{1}{2}\sum_{i,j=1}^{l+u} w_{i,j}(f(x_i) - f(x_j))^2 = \frac{1}{2}f(X)^T L_a f(X), \tag{6.1}$$

where $f(X) = [f(x_1), f(x_2), \ldots, f(x_{l+u})]$ represents the decision function values over all the training data X, $L_a = D - W$ is the graph Laplacian matrix, $W = (w_{i,j})$ is the adjacency matrix of dimension $\{(l+u) \times (l+u)\}$ and its $w_{i,j}$ entry corresponds to edge-weight defined for a pair of points (x_i, x_j), D is the diagonal matrix given by $D_{i,i} = \sum_{j=1}^{l+u} w_{i,j}$.

With a selection of kernel function $K(\cdot, \cdot)$ with norm $||\cdot||_H$ in the Reproducing Kernel Hilbert Space (RKHS), the semi-supervised manifold regularization framework is established by Minimizing

$$f^* = arg \underset{(f \in H)}{\text{Min}} \ (R^{emp}(f) + \gamma_H ||f||_H^2 + \gamma_M ||f||_M^2), \tag{6.2}$$

where $R^{emp}(f)$ denotes the empirical risk on the labeled data X_l, γ_H is the parameter corresponding to $||f||_H^2$ which penalizes the complexity of f in the RKHS, and γ_M is the parameter associated with $||f||_M^2$ which enforces smoothness of function 'f' along the intrinsic manifold M. For more details refer Belkin et al. [1].

6.2 Laplacian-SVM

Laplacian-SVM(Lap-SVM) follows the principles behind manifold regularization (Belkin et al. [1]). In order to learn a Lap-SVM classifier, the following problem is solved

$$\underset{(f \in H_k)}{\text{Min}} \sum_{i=1}^{l} Max(1 - y_i f(x_i), 0) + \gamma_H ||f||_H^2 + \gamma_M ||f||_M^2. \tag{6.3}$$

By introducing the slack variables ξ_i, the above unconstrained primal problem can be written as a constrained optimization problem, with decision variable (α, b, ξ) as

$$\begin{aligned}
\underset{(\alpha,b,\xi)}{\text{Min}} \quad & \sum_{i=1}^{l} \xi_i + \gamma_H \alpha^T K \alpha + \gamma_M \alpha^T K L K \alpha \\
\text{subject to} \quad & \\
& y_i \left(\textstyle\sum_{j=1}^{n} \alpha_i K(x_i, x_j) + b\right) \geq 1 - \xi_i, \qquad (i = 1, 2, \ldots, l), \\
& \xi_i \geq 0, \qquad (i = 1, 2, \ldots, l).
\end{aligned}$$

The Wolfe dual of the optimization problem (6.2) can be written as

$$\begin{aligned}
\underset{\beta}{\text{Max}} \quad & \sum_{i=1}^{l} \beta_i - \frac{1}{2} \beta^T Q \beta \\
\text{subject to} \quad & \\
& \sum_{i=1}^{l} \beta_i y_i = 0, \\
& 0 \leq \beta_i \leq 1, \qquad (i = 1, 2, \ldots, l),
\end{aligned} \tag{6.4}$$

where,
$Q = YJ_LK(2\gamma_H I + 2\gamma_M KL)^{-1}J_L^T Y$, $Y \in R^{l,l}$, $J_L \in R^{1,l}$ and $I \in R^{l,l}$ is the identity matrix where $R^{l,l}$ is a matrix of order $l \times l$.

Once the Lagrange multiplier β is obtained, the optimal value of decision variable α^* is determined with

$$\alpha^* = (2\gamma_H I + 2\gamma_M KL)^{-1}J_L^T Y\beta.$$

The target function f^* is defined as $f^*(x) = \sum_{i=1}^{l} \alpha_i^* K(x_i, x)$.

The decision function that discriminates between class $+1$ and -1 is given by $y(x) = sign(f^*(x))$.

6.3 Laplacian-TWSVM

Similar to SVM, the decision hyperplanes corresponding to Laplacian TWSVM (Lap-TWSVM) can be written as

$$f_1(x) = (w_1 \cdot x) + b_1, \tag{6.5}$$

and

$$f_2(x) = (w_2 \cdot x) + b_2, \tag{6.6}$$

where $[w_1, b_1]$ and $[w_2, b_2]$ are decision variables corresponding to Lap-TWSVM hyperplanes. In similar to the TWSVM, authors in [10] have used the combination of square loss function and hinge loss function for Laplacian-TWSVM. Thus, the loss function for Laplacian-TWSVM (Lap-TWSVM) can be expressed as

$$V_1(x_i, y_i, f_1) = ((A_i \cdot w_1) + b_1)^2 + Max(0, 1 - f_1(B_i)), \tag{6.7}$$
$$V_2(x_i, y_i, f_2) = ((B_i \cdot w_2) + b_2)^2 + Max(0, 1 - f_2(A_i)). \tag{6.8}$$

where A_i and B_i are i^{th} row of matrices A (matrix corresponding to class label $+1$) and B (matrix corresponding to class label -1) and (w_1, b_1) and (w_2, b_2) are augmented matrix or vector corresponding to classes A and B respectively.

On the lines of Lap-SVM, the regularization terms of Lap-TWSVM i.e. $||f_1||_H^2$ and $||f_2||_H^2$ can be expressed as

$$||f_1||_H^2 = \frac{1}{2}(||w_1||_2^2 + b_1^2), \tag{6.9}$$

and

$$||f_2||_H^2 = \frac{1}{2}(||w_2||_2^2 + b_2^2), \tag{6.10}$$

respectively.

For manifold regularization, a data adjacency graph $W_{(l+u)x(l+u)}$ is defined by nodes $W_{i,j}$, which represents the similarity of every pair of input samples. The weight matrix W may be defined by k nearest neighbor as follows

$$W_{ij} = \begin{cases} exp(-||x_i - x_j||_2^2/2\sigma^2), & \text{if } x_i, x_j \text{ are neighbor,} \\ 0, & \text{Otherwise,} \end{cases} \tag{6.11}$$

where $\sigma > 0$ is a parameter.

Thus, the manifold regularization is defined by

$$||f_1||_M^2 = \frac{1}{(l+u)^2} \sum_{i,j=1}^{l+u} W_{i,j}(f_1(x_i) - f_1(x_j))^2 = f_1^T L f_1, \tag{6.12}$$

$$||f_2||_M^2 = \frac{1}{(l+u)^2} \sum_{i,j=1}^{l+u} W_{i,j}(f_2(x_i) - f_2(x_j))^2 = f_2^T L f_2, \tag{6.13}$$

where $L = D - W$ is the graph Laplacian, D is a degree matrix, $f_1 = [f_1(x_1), \ldots, f_1(x_{l+u})]^T = Xw_1 + eb_1, f_2 = [f_2(x_1), \ldots, f_2(x_{l+u})]^T = Xw_2 + eb_2$, e is an appropriate ones vector, and X is the data matrix consisting of all the samples.

With these definitions, the primal problem of Lap-TWSVM is defined as

$$\begin{aligned} \underset{(w_1, b_1, \xi_2)}{\text{Min}} \quad & \frac{1}{2}||Aw_1 + e_1 b_1||^2 + c_1 e_2^T \xi_2 + \frac{c_2}{2}(w_1^T w_1 + b_1) \\ & + \frac{c_3}{2}(w_1^T X + e^T b_1) L (Xw_1 + eb_1) \end{aligned}$$

subject to

$$\begin{aligned} -(Bw_1 + e_2 b_1) + \xi_2 &\geq e_2, \\ \xi_2 &\geq 0, \end{aligned}$$

and

$$\begin{aligned} \underset{(w_2, b_2, \xi_1)}{\text{Min}} \quad & \tfrac{1}{2}||Bw_2 + e_2 b_2||^2 + c_1 e_1^T \xi_1 + \frac{c_2}{2}(w_2^T w_2 + b_2) \\ & + \frac{c_3}{2}(w_2^T X + e^T b_2) L (Xw_2 + eb_2) \end{aligned}$$

subject to

$$\begin{aligned} (Aw_2 + e_1 b_2) + \xi_1 &\geq e_1, \\ \xi_1 &\geq 0, \end{aligned}$$

where e_1 and e_2 are the vector of ones of dimensions equal to number of known label pattern in the respective classes, c_1 & $c_2 > 0$ are trade-off parameter.

Similarly, for nonlinear kernel the corresponding decision function with parameter (λ_1, b_1) and (λ_2, b_2) can be expressed as the kernel-generated hyperplanes given by

$$K(x^T, X^T)\lambda_1 + b_1 = 0, \tag{6.14}$$
$$K(x^T, X^T)\lambda_2 + b_2 = 0, \tag{6.15}$$

where K is chosen kernel function: $K(x_i, x_j) = (\phi(x_i) \cdot \phi(x_j))$, X is the data matrix comprising of both labelled as well as unlabelled patterns. With the help of above notations the regularizer term can be expressed as

$$||f_1||_H^2 = \frac{1}{2}(\lambda_1^T K \lambda_1 + b_1^2), \tag{6.16}$$
$$||f_2||_H^2 = \frac{1}{2}(\lambda_2^T K \lambda_2 + b_2^2). \tag{6.17}$$

Similar to linear case, for manifold regularization, $||f_1||_M^2$ and $||f_2||_M^2$ can be written as

$$||f_1||_M^2 = f_1^T L f_1 = (\lambda_1^T K + e^T b_1) L (K\lambda_1 + e b_1), \tag{6.18}$$
$$||f_2||_M^2 = f_2^T L f_2 = (\lambda_2^T K + e^T b_2) L (K\lambda_2 + e b_2), \tag{6.19}$$

where 'e' is vector of ones of appropriate dimensions.

Thus, the nonlinear optimization problem as discussed in [10]

$$\begin{aligned} \underset{(\lambda_1, b_1, \xi_2)}{\text{Min}} \quad & \tfrac{1}{2}||K(A, X^T)\lambda_1 + e_1 b_1||^2 + c_1 e_2^T \xi_2 + \frac{c_2}{2}(\lambda_1^T K \lambda_1 + b_1^2) \\ & + \frac{c_3}{2}(\lambda_1^T K + e^T b_1) L (K\lambda_1 + e b_1) \end{aligned}$$

subject to

$$\begin{aligned} -(K(B, X^T)\lambda_1 + e_2 b_1) + \xi_2 &\geq e_2, \\ \xi_2 &\geq 0, \end{aligned}$$

and

$$\begin{aligned} \underset{(\lambda_2, b_2, \xi_1)}{\text{Min}} \quad & \frac{1}{2}||K(B, X^T)\lambda_2 + e_2 b_2||^2 + c_1 e_1^T \xi_1 + \frac{c_2}{2}(\lambda_2^T K \lambda_2 + b_2^2) \\ & + \frac{c_3}{2}(\lambda_2^T K + e^T b_2) L (K\lambda_2 + e b_2) \end{aligned}$$

subject to

$$\begin{aligned} (K(A, X^T)\lambda_2 + e_1 b_2) + \xi_1 &\geq e_1, \\ \xi_1 &\geq 0, \end{aligned} \tag{6.20}$$

where e_1 and e_2 are the vector of ones of dimensions equal to number of known label pattern in respective classes. Similar to c_1 and c_2, $c_3 > 0$ is trade-off parameter.

From Qi et al. [10], the Wolfe dual problem corresponding to hyperplane class +1 can be expressed as

$$\begin{array}{ll} \underset{\alpha}{\text{Max}} & e_2^T\alpha - \frac{1}{2}(\alpha^T G_\phi)(H_\phi^T H_\phi + c_2 O_\phi + c_3 J_\phi^T L J_\phi)^{-1}(G_\phi^T \alpha) \\ \text{subject to} & \end{array}$$

$$0 \leq \alpha \leq c_1 e_2, \tag{6.21}$$

where,

$$H_\phi = \begin{bmatrix} K(A, X^T) & e_1 \end{bmatrix}, \quad O_\phi = \begin{bmatrix} K & 0 \\ 0 & 0 \end{bmatrix}, \quad J_\phi = \begin{bmatrix} K & e \end{bmatrix}, \quad G_\phi = \begin{bmatrix} K(B, X^T) & e_2 \end{bmatrix},$$

and the augmented vector $u_1 = [\lambda_1 \;\; b_1]^T$.

Once the vector of dual variables α is determined the vector of primal variables is obtained by solving following equations:

$$u_1 = -(H_\phi^T H_\phi + c_2 O_\phi + c_3 J_\phi^T L J_\phi)^{-1}(G^T \alpha). \tag{6.22}$$

In a similar manner, the Wolf dual problem corresponding to hyperplane class -1 can be expressed as

$$\begin{array}{ll} \underset{\beta}{\text{Max}} & e_1^T\beta - \frac{1}{2}(\beta^T P_\phi)(Q_\phi^T Q_\phi + c_2 U_\phi + c_3 F_\phi^T L F_\phi)^{-1}(P_\phi^T \alpha) \\ \text{subject to} & \end{array}$$

$$0 \leq \beta \leq c_2 e_1, \tag{6.23}$$

where,

$$Q_\phi = \begin{bmatrix} K(B, X^T) & e_2 \end{bmatrix}, \quad U_\phi = \begin{bmatrix} K & 0 \\ 0 & 0 \end{bmatrix} \quad F_\phi = \begin{bmatrix} K & e \end{bmatrix}, \quad P_\phi = \begin{bmatrix} K(A, X^T) & e_1 \end{bmatrix}$$

and the augmented vector $u_2 = [\lambda_2 \;\; b_2]^T$.

Similar to the above problem, the primal variable u_2 is obtained as

$$u_2 = (Q_\phi^T Q_\phi + c_2 U_\phi + c_3 F_\phi^T L F_\phi)^{-1}(P^T \beta). \tag{6.24}$$

Once the vectors u_1 and u_2 are obtained from above equations, a new data point $x \in R^n$ is then assigned to the +1 and -1 class, based on which of the two hyperplanes it lies closest to, i.e.

$$f(x) = arg \underset{(i=1,2)}{\text{Min}} \; d_i(x), \tag{6.25}$$

where,

$$d_i = \left| \frac{\lambda_i^T K(x^T, X^T) + b_i}{||\lambda_i||} \right|, \tag{6.26}$$

where $|\cdot|$ is the perpendicular distance of point x from the decisions hyperplane.

6.4 Laplacian Least Squares TWSVM

Taking motivation from Least Squares SVM (LSSVM) [7] and proximal SVM (PSVM) [12], authors in [11] elaborated the formulation of Laplacian Least Squares Twin SVM for binary data classification. Later on, Khemchandani et al. [13] extended Laplacian Least Squares TWSVM for classifying multicategory datasets with One versus One versus Rest strategy. In [11], authors modify the loss function as

$$R_1^{emp} = \sum_{i=1}^{m_1} f_1(A_i)^2 + c_1 \sum_{i=1}^{m_2} (f_1(B_i) + 1)^2, \tag{6.27}$$

and

$$R_2^{emp} = \sum_{i=1}^{m_2} f_2(B_i)^2 + c_2 \sum_{i=1}^{m_1} (f_2(A_i) - 1)^2, \tag{6.28}$$

where $f_1(A_i)$ and $f_2(B_i)$ represent decision function values over the training data belonging to class A and B respectively and $c_1 > 0$ & $c_2 > 0$, are the risk penalty parameter which determine the trade-off between the loss terms in Eqs. (6.27) and (6.28) respectively.

By introducing the slack variables (error variables) y_1, y_2, z_1 and z_2 of appropriate dimensions, in the corresponding primal problem of Lap-LSTWSVM, we obtain the following optimization problems,

$$\underset{(w_1,b_1,y_1,z_1)}{\text{Min}} \quad \frac{1}{2}(y_1^T y_1 + c_1 z_1^T z_1) + \frac{1-\lambda_1}{2}(Xw_1 + eb_1)^T L_a (Xw_1 + eb_1) + \frac{\lambda_1}{2}(||w_1||^2 + b_1^2)$$

subject to

$$\begin{aligned} Aw_1 + e_1 b_1 &= y_1, \\ Bw_1 + e_2 b_1 + e_2 &= z_1, \end{aligned} \tag{6.29}$$

and

$$\underset{(w_2,b_2,y_2,z_2)}{\text{Min}} \quad \frac{1}{2}(z_2^T z_2 + c_2 y_2^T y_2) + \frac{1-\lambda_2}{2}(Xw_2 + eb_2)^T L_a (Xw_2 + eb_2) + \frac{\lambda_2}{2}(||w_2||^2 + b_2^2)$$

subject to

$$\begin{aligned} Bw_2 + e_2 b_2 &= z_2, \\ Aw_2 + e_1 b_2 - e_1 &= y_2, \end{aligned} \tag{6.30}$$

where $\lambda_1, \lambda_2 \in (0, 1]$ are the regularization parameters and L_a is corresponding Laplacian matrix and e_1, e_2 are vector of ones of appropriate dimensions. Substituting the equality constraints into the objective function leads to the following optimization problem,

$$\underset{(w_1,b_1)}{\text{Min}} \quad L = \frac{1}{2}(||Aw_1 + e_1 b_1||^2 + c_1||Bw_1 + e_2 b_1 + e_2||^2) + \frac{\lambda_1}{2}(||w_1||^2 + b_1^2) + \frac{1-\lambda_1}{2}(Xw_1 + eb_1)^T L_a (Xw_1 + eb_1) \tag{6.31}$$

Equating the partial derivative of Eq. (6.31) with respect to w_1 and b_1 equal to zero leads to following set of equations,

$$\begin{aligned} \nabla_{w_1} L = {} & A^T(Aw_1 + e_1 b_1) + c_1 B^T(Bw_1 + e_2 b_1 + e_2) \\ & + \lambda_1 w_1 + (1-\lambda_1)X^T L_a (Xw_1 + eb_1) = 0, \end{aligned} \tag{6.32}$$

$$\begin{aligned} \nabla_{b_1} L = {} & e_1^T(Aw_1 + e_1 b_1) + c_1 e_2^T(Bw_1 + e_2 b_1 + e_2) \\ & + \lambda_1 b_1 + (1-\lambda_1)e^T L_a (Xw_1 + eb_1) = 0. \end{aligned} \tag{6.33}$$

Combining Eqs. (6.32) and (6.33) in matrix form, we get

$$\begin{bmatrix} A^T \\ e_1^T \end{bmatrix} \begin{bmatrix} A & e_1 \end{bmatrix} \begin{bmatrix} w_1 \\ b_1 \end{bmatrix} + c_1 \begin{bmatrix} B^T \\ e_2^T \end{bmatrix} \begin{bmatrix} B & e_2 \end{bmatrix} \begin{bmatrix} w_1 \\ b_1 \end{bmatrix} + c_1 \begin{bmatrix} B^T \\ e_2^T \end{bmatrix} + \lambda_1 \begin{bmatrix} w_1 \\ b_1 \end{bmatrix} + (1-\lambda_1) \begin{bmatrix} X^T \\ e^T \end{bmatrix} L_a \begin{bmatrix} X & e \end{bmatrix} \begin{bmatrix} w_1 \\ b_1 \end{bmatrix} = 0. \tag{6.34}$$

Let $J = \begin{bmatrix} X & e \end{bmatrix}$, $H = \begin{bmatrix} A & e_1 \end{bmatrix}$, $G = \begin{bmatrix} B & e_2 \end{bmatrix}$, $v_1 = \begin{bmatrix} w_1^T & b_1 \end{bmatrix}^T$ and solving Eq. (6.34) for w_1 and b_1, is equivalent to solving following system of linear equations are

$$Pv_1 = -c_1 G^T e_2, \tag{6.35}$$

where $P = H^T H + c_1 G^T G + \lambda_1 I + (1-\lambda_1)J^T L_a J$.

In a similar way, the solution of Eq. (6.30) could be obtained by solving following system of linear equations,

$$Qv_2 = c_2 H^T e_1, \tag{6.36}$$

where, $v_2 = [w_2^T \;\; b_2]^T$ and $Q = G^T G + c_2 H^T H + \lambda_2 I + (1 - \lambda_2) J^T L_a J$.

Once the solution (w_1, b_1) and (w_2, b_2) of the optimization problem is obtained, a new data $x \in R^n$ is assigned to class i ("$i = +1$" or "-1"), depending on which of the two hyperplanes it is closer to, i.e.

$$Class\ i = arg \underset{(k=1,2)}{\text{Min}} \frac{|w_k^T x + b_k|}{|w_k|}, \tag{6.37}$$

where $|z|$ is the absolute value of z.

6.5 Unsupervised Learning

Clustering is a powerful tool which aims at grouping similar objects into the same cluster and dissimilar objects into different clusters by identifying dominant structures in the data. It has remained a widely studied research area in machine learning (Anderberg [14]; Jain et al. [15]; Aldenderfer and Blashfield [16]) and has applications in diverse domains such as computer vision, text mining, bioinformatics and signal processing (QiMin et al. [17]; Zhan et al. [18], Tu et al. [19]; Liu and Li [20]).

Traditional point based clustering methods such as k-means (Anderberg [14]), k-median (Bradley and Mangasarian [21]) etc. work by partitioning the data into clusters based on the cluster prototype points. These methods perform poorly in case when data is not distributed around several cluster points. In contrast to these, plane based clustering methods such as k-plane clustering (Bradley and Mangasarian [22]), proximal plane clustering ([23]), local k-proximal plane clustering (Yang et. al. [24]) etc. have been proposed in literature. These methods calculate k cluster center planes and partition the data into k clusters according to the proximity of the datapoints with these k planes.

Recently, Wang et al. [25] proposed a novel plane based clustering method namely twin support vector clustering (TWSVC). The method is based on twin support vector machine (TWSVM) (Jayadeva et al. [6]) and exploits information from both within and between clusters. Different from the TWSVM, the formulation of TWSVC is modified to get one cluster plane close to the points of its own cluster and at the same time far away from the points of different clusters from both sides of cluster planes. Experiment results (Wang et al. [25]) show the superiority of the method against existing plane based methods.

The samples are denoted by a set of m row vectors $X = \{x_1; x_2; \ldots; x_m\}$ in the n-dimensional real space $\mathbb{R}^n$, where the jth sample $x_j = (x_{j1}, x_{j2}, \ldots, x_{jn})$. We assume that these samples belong to k clusters with their corresponding cluster labels in

$\{1, 2, \ldots, k\}$. Let X_i denotes the set of samples belonging to cluster label i and $\overline{X_i}$, where $i = 1, 2, \ldots, k$ denotes the set of samples belonging to other than i cluster label.

6.5.1 *k-Means*

Consider the clustering problem with a set X of m unlabeled data samples in $\mathbb{R}^n$. The method of k-means (Anderberg [14]) partitions X into k clusters $X_1, X_2, \ldots, X_k$ such that the data samples are close to their respective k cluster center points $\mu_1, \mu_2, \ldots, \mu_k$. It aims to minimize the following objective function

$$\underset{(\mu_1,\mu_2,\ldots,\mu_k,X_1,X_2,\ldots,X_k)}{\text{Min}} \quad \sum_{i=1}^{k}\sum_{j=1}^{m_i}\|X_i(j) - \mu_i\|_2, \tag{6.38}$$

where $X_i(j)$ represents the jth sample in X_i, m_i is the number of samples in X_i so that $m_1 + m_2 + \cdots + m_k = m$, and $\|.\|_2$ denotes L_2 norm.

In practice, an iterative relocation algorithm is followed which minimizes (6.38) locally. Given an initial set of k cluster center points, each sample x is labelled to its nearest cluster center by

$$y = \underset{i}{arg \;\; \text{Min}} \quad \{\|x - \mu_i\|, i = 1, 2, \ldots, k\}. \tag{6.39}$$

Then the k cluster center points are updated as the mean of the corresponding cluster samples since for a given assignment X_i, the mean of the cluster samples represents the solution to (6.38). At each iteration, the cluster centers and sample labels are updated until some convergence criteria is satisfied.

6.5.2 *TWSVC*

Considers the following problem by Wang et al. [25] to get k cluster center planes $w_i^T x + b_i = 0, i = 1, 2, \ldots, k$, one for each class

$$\begin{aligned}
&\underset{(w_i,\, b_i,\, q_i, X_i)}{\text{Min}} \quad \frac{1}{2}\|(X_i w_i + b_i e)\|_2 + Ce^T q_i \\
&\text{subject to} \\
&\qquad |(\overline{X_i} w_i + b_i e)| + q_i \geq e, \\
&\qquad\qquad q_i \geq 0,
\end{aligned} \tag{6.40}$$

where $C > 0$ is a penalty parameter and q_i is a slack vector.

Each of the k hyperplane is close to the samples of its cluster and far away from the samples of the other clusters from both sides unlike the multi class TWSVM which yields hyperplanes which are close to the samples of its class but are away from the samples of other class from one side only.

Wang et al. [25] solves Eq. (6.40) for a given X_i by the concave-convex procedure (CCCP) ([26]), which decomposes it into a series of convex quadratic subproblems with an initial w_i^0 and b_i^0 as follows:

$$\begin{aligned}
&\underset{(w_i^{j+1},\, b_i^{j+1},\, q_i^{j+1})}{\text{Min}} \quad \frac{1}{2}\|(X_i w_i^{j+1} + b_i^{j+1} e)\|_2 + Ce^T q_i^{j+1} \\
&\text{subject to} \\
&\qquad T(|(\overline{X}_i w_i^{j+1} + b_i^{j+1} e)|) + q_i^{j+1} \geq e, \\
&\qquad q_i^{j+1} \geq 0,
\end{aligned} \tag{6.41}$$

where the index of the subproblem $j = 0, 1, 2, \ldots,$ and $T(.)$ denotes the first order Taylor expansion.

Wang et al. [25] shows that the above problem (6.41) becomes equivalent to the following:

$$\begin{aligned}
&\underset{(w_i^{j+1},\, b_i^{j+1},\, q_i^{j+1})}{\text{Min}} \quad \tfrac{1}{2}\|(X_i w_i^{j+1} + b_i^{j+1} e)\|_2 + Ce^T q_i^{j+1} \\
&\text{subject to} \\
&\qquad diag(sign(\overline{X}_i w_i^j + b_i^j e))(\overline{X}_i w_i^{j+1} + b_i^{j+1} e) + q_i^{j+1} \geq e, \\
&\qquad q_i^{j+1} \geq 0,
\end{aligned} \tag{6.42}$$

which is solved for $[w_i^{j+1}; b_i^{j+1}]$ by solving its dual problem

$$\begin{aligned}
&\underset{\alpha}{\text{Min}} \ \frac{1}{2}\alpha^T G(H^T H)^{-1} G^T \alpha - e^T \alpha \\
&\text{subject to} \\
&\qquad 0 \leq \alpha \leq Ce,
\end{aligned} \tag{6.43}$$

where $G = diag(sign(\overline{X}_i w_i^j + b_i^j e))[\overline{X}_i \quad e]$, $H = [X_i \quad e]$ and $\alpha \in \mathbb{R}^{m-m_i}$ is the Lagrangian multiplier vector.

The solution to (6.42) is obtained from the solution to (6.43) by

$$[w_i^{j+1}; b_i^{j+1}]^T = (H^T H)^{-1} G^T \alpha, \tag{6.44}$$

In short, for each $i = 1, 2, \ldots, k$, we select an initial w_i^0 and b_i^0 and solve for $[w_i^{j+1}; b_i^{j+1}]$ by (6.44) for $j = 0, 1, 2 \ldots,$ and stop when $\|[w_i^{j+1}; b_i^{j+1}] - [w_i^j; b_i^j]\|$ is small enough. We then set $w_i = w_i^{j+1}$ and $b_i = b_i^{j+1}$.

Given any initial sample cluster assignment of X, TWSVC iterates alternately updating the cluster center planes by solving (6.40) with a certain X_i and then update cluster assignments by relabelling each sample by $y = \underset{i}{arg\ Min}\{|w_i^T x + b_i|, i = 1, 2, \ldots, k\}$. The iterations are repeated until some convergence criteria is met.

It is to be noted that the solution of Eq. (6.42) requires solving a QPP with $m - m_i$ parameters and in addition requires an inversion of matrix of size $(n+1) \times (n+1)$ where $n \ll m$.

TWSVC was also extended in Wang et al. [25] to handle nonlinear case by considering k cluster center kernel generated surfaces for $i = 1, 2, \ldots, k$

$$K(x, X)u_i + \gamma_i = 0, \tag{6.45}$$

where K is any arbitrary kernel, $u_i \in \mathbb{R}^m$ and $\gamma \in \mathbb{R}$. The kernel counterpart of (6.40) for $i = 1, 2, \ldots, k$ is

$$\begin{aligned} &\underset{(u_i,\, \gamma_i,\, \eta_i, X_i)}{\text{Min}} \quad \frac{1}{2}\|(K(X_i, X)u_i + \gamma_i e)\|_2 + Ce^T\eta_i \\ &\text{subject to} \\ &\qquad |(K(\overline{X_i}, X)u_i + \gamma_i e)| + \eta_i \geq e, \\ &\qquad\qquad\qquad\qquad \eta_i \geq 0, \end{aligned} \tag{6.46}$$

where η_i is a error vector. The above problem is solved in a similar manner to linear case by CCCP. However, it is worth mentioning that for each $i = 1, 2, \ldots, k$, the solution of nonlinear TWSVC is decomposed into solving a series of subproblems which requires inversion of matrix of size $(m+1) \times (m+1)$ along with a QPP to be solved, where m is the total number of patterns.

6.6 Fuzzy Least Squares Twin Support Vector Clustering

Taking motivation from Kumar and Gopal [8], Khemchandani et al. in [27] proposed Least Squares version of TWSVC and then extend it to fuzzy LS-TWSVC. Here, we modify the primal problem of linear TWSVC (6.40) in least squares sense (Suykens and Vandewalle [7]), with inequality constraints replaced with equality constraints along with adding a regularization term in the objective function to incorporate Structural Risk Minimization (SRM) principle. Thus, for class i $(i = 1, 2, \ldots, k)$ the optimization problem is given as:

$$\begin{aligned} &\underset{(w_i,\, b_i,\, q_i, X_i)}{\text{Min}} \quad \frac{1}{2}\|(X_i w_i + b_i e)\|_2 + \frac{\nu}{2}(\|w_i\|_2 + b_i^2) + \frac{C}{2}\|q_i\|_2 \\ &\text{subject to} \end{aligned}$$

$$|(\overline{X_i}w_i + b_ie)| + q_i = e, \tag{6.47}$$

where $C > 0$, $\nu > 0$ are trade-off parameter. Note that QPP (6.47) uses the square of L2-norm of slack variable q_i instead of 1-norm of q_i in (6.40), which makes the constraint $q_i \geq 0$ redundant (Fung and Mangasarian [12]). Further solving (6.47) is equivalent to solving system of linear equations.

Further, we introduce the fuzzy matrices S_i and $\overline{S_i}$ in (6.47) which indicates the fuzzy membership value of each data point to different available clusters as follows

$$\underset{(w_i,\, b_i,\, q_i, X_i)}{\text{Min}} \quad \frac{1}{2}\|(S_iX_iw_i + b_ie)\|_2 + \frac{\nu}{2}(\|w_i\|_2 + b_i^2) + \frac{C}{2}\|q_i\|_2$$
subject to

$$|(\overline{S_iX_i}w_i + b_ie)| + q_i = e. \tag{6.48}$$

Similar to the solution of TWSVC formulation (Wang et al. [25]), the above optimization problem can be solved by using the concave-convex procedure (CCCP) (Yuille and Rangarajan [26]), which decomposes it into a series of j $(j = 0, 1, 2, \ldots)$ quadratic subproblems with initial w_i^0 and b_i^0 as follows

$$\underset{(w_i^{j+1},\, b_i^{j+1},\, q_i^{j+1})}{\text{Min}} \quad \frac{1}{2}(\|(S_iX_iw_i^{j+1} + b_i^{j+1}e)\|_2) + \frac{\nu}{2}(\|w_i^{j+1}\|_2 + (b_i^{j+1})^2) + \frac{C}{2}\|q_i^{j+1}\|_2$$
subject to

$$T(|(\overline{S_iX_i}w_i^{j+1} + b_i^{j+1}e)|) + q_i^{j+1} = e, \tag{6.49}$$

where $T(.)$ denotes the first order Taylor expansion.

Working along the lines of Wang et al. [25], the Eq. (6.49) reduces to

$$\underset{(w_i^{j+1},\, b_i^{j+1},\, q_i^{j+1})}{\text{Min}} \quad \frac{1}{2}(\|(S_iX_iw_i^{j+1} + b_i^{j+1}e)\|_2) + \frac{\nu}{2}(\|w_i^{j+1}\|_2 + (b_i^{j+1})^2) + \frac{C}{2}\|q_i^{j+1}\|_2$$
subject to

$$diag(sign(\overline{S_iX_i}w_i^j + b_i^je))(\overline{S_iX_i}w_i^{j+1} + b_i^{j+1}e) + q_i^{j+1} = e. \tag{6.50}$$

Substituting the error variable q_i^{j+1} into the objective function of (6.50) leads to the following optimization problem.

$$\underset{(w_i^{j+1},\, b_i^{j+1})}{\text{Min}} \quad \frac{1}{2}(\|(S_iX_iw_i^{j+1} + b_i^{j+1}e)\|_2) + \frac{\nu}{2}(\|w_i^{j+1}\|_2 + (b_i^{j+1})^2) +$$
$$\frac{C}{2}\|diag(sign(\overline{S_iX_i}w_i^j + b_i^je))(\overline{S_iX_i}w_i^{j+1} + b_i^{j+1}e) - e\|_2. \tag{6.51}$$

Further, considering the gradient of (6.51) with respect to w_i^{j+1} and b_i^{j+1} and equate it to zero gives:

$$(S_iX_i)^T[H_1z_i^{j+1}] + \nu w_i^{j+1} + C(\overline{S_iX_i})^T G^T[G(H_2z_i^{j+1}) - e] = 0, \quad (6.52)$$

$$e^T[H_1z_i^{j+1}] + \nu b_i^{j+1} + Ce^T G^T[G(H_2z_i^{j+1}) - e] = 0, \quad (6.53)$$

where $H_1 = [S_iX_i \quad e]$, $H_2 = [\overline{S_iX_i} \quad e]$, $z_i^{j+1} = [w_i^{j+1}; b_i^{j+1}]$ and $G = diag(sign(H_2 \ z_i^j))$. Rearranging the above equations, we obtained the following system of linear equations:

$$(H_1^T H_1 + I\nu + CH_2^T H_2)z_i^{j+1} = CH_2^T G^T e, \quad (6.54)$$

which gives the solution for z_i^{j+1}:

$$z_i^{j+1} = [w_i^{j+1}; b_i^{j+1}] = C(H_1^T H_1 + I\nu + CH_2^T H_2)^{-1} H_2^T G^T e. \quad (6.55)$$

It can be finally observed that our algorithm requires the solution of (6.55) which involves inversion of smaller dimensional matrix of size $(n+1) \times (n+1)$ as compared to an additional QPP solution required in case of TWSVC.

6.7 Nonlinear Fuzzy Least Squares Twin Support Vector Clustering

Working on the lines of Jayadeva et al. [6], we extend the nonlinear formulation of LS-TWSVC and F-LS-TWSVC by considering k cluster center kernel generated surfaces for $i = 1, 2, \ldots, k$:

$$K(x, X)u_i + \gamma_i = 0, \quad (6.56)$$

where K is any arbitrary kernel, $u_i \in \mathbb{R}^m$ and $\gamma_i \in \mathbb{R}$. The primal QPP of LS-TWSVC (6.46) is modified in least squares sense as follows for $i = 1, 2, \ldots, k$:

$$\underset{(u_i,\, \gamma_i,\, \eta_i)}{\text{Min}} \ \frac{1}{2}(\|(K(X_i, X)u_i + \gamma_i e)\|_2) + \frac{\nu}{2}(u_i^2 + \gamma_i^2) + C\eta_i^T \eta_i$$

subject to

$$(|(K(\overline{X_i}, X)u_i + \gamma_i e)|) + \eta_i = e. \quad (6.57)$$

Similar to the linear case, for each $i = 1, 2, \ldots, k$ the above problem is also decomposed into series of quadratic subproblems where the index of subproblems is $j = 0, 1, 2 \ldots,$ and solution of which can be derived to be:

$$[u_i^{j+1}; \gamma_i^{j+1}] = C(E_1^T E_1 + I\nu + CE_2^T E_2)^{-1} E_2^T F^T e, \tag{6.58}$$

where $E_1 = [(K(X_i, X)) \;\; e]$, $E_2 = [(K(\overline{X_i}, X)) \;\; e]$ and $F = diag(sign (E_2[u_i^j; b_i^j]))$.

Further, we introduce the fuzzy matrices S_i and $\overline{S_i}$ in (6.57) which indicates the fuzzy membership value of each data point to different clusters. As a result, the primal QPP of F-LS-TWSVC can be written as follows for $i = 1, 2, \ldots, k$

$$\underset{(u_i,\, \gamma_i,\, \eta_i)}{\text{Min}} \quad \frac{1}{2}(\|(S_i K(X_i, X)u_i + \gamma_i e)\|_2) + \frac{\nu}{2}(u_i^2 + \gamma_i^2) + C\eta_i^T \eta_i$$
subject to

$$(|(\overline{S_i} K(\overline{X_i}, X)u_i + \gamma_i e)|) + \eta_i = e. \tag{6.59}$$

For each $i = 1, 2, \ldots, k$ the above problem is also decomposed into series of quadratic subproblems where the index of subproblems is $j = 0, 1, 2 \ldots,$ and solution of which can be derived to be

$$[u_i^{j+1}; \gamma_i^{j+1}] = C(E_1^T E_1 + I\nu + CE_2^T E_2)^{-1} E_2^T F^T e, \tag{6.60}$$

where $E_1 = [S_i(K(X_i, X)) \;\; e]$, $E_2 = [\overline{S_i}(K(\overline{X_i}, X)) \;\; e]$ and $F = diag(sign (E_2[u_i^j; b_i^j]))$.

The overall algorithm remains the same as of linear case except that we solve for k kernel generated surfaces parameters $u_i, \gamma_i, i = 1, 2, \ldots, k$.

It can be noted that the nonlinear algorithm requires the solution of (6.60) which involves calculating the inverse of matrix of order $(m + 1) \times (m + 1)$. However, inversion in (6.60) can be solved by calculating inverses of two smaller dimension matrices as compare to $(m + 1) \times (m + 1)$ by using Sherman-Morrison-Woodbury (SMW) (Golub and Van Loan [28]) formula. Therefore, inversion of matrices in (6.60) can be further solved by

$$[u_i^{j+1}; \gamma_i^{j+1}] = C(Y - YE_1^T(I + E_1 Y E_1^T)^{-1} E_1 Y) E_2^T F^T e, \tag{6.61}$$

where $Y = \frac{1}{\nu}(I - E_2^T(\frac{\nu}{C} + E_2 E_2^T)^{-1} E_2)$, which involves matrix inverses of $(m_i \times m_i)$ and $((m - m_i) \times (m - m_i))$ respectively for $i = 1, 2, \ldots, k$.

Input : The dataset X; the number of clusters k; appropriate F-LS-TWSVC parameters C, ν.
Output: k fuzzy matrices S^i for $i = 1, 2, ..., k$
Process:
1. Initialize fuzzy membership matrix S via FNNG (as explained in Sect. 6.8.3) for each data points in k clusters.
2. For each $i = 1, 2, ..., k$:
 2.1. Use obtained fuzzy membership matrix in Step 1 as initial fuzzy membership matrix S_0^j and solve Eq. (6.55) to obtain $[w_i^{j+1} \;\; b_i^{j+1}]$ and $S_i^{j+1} = \frac{1}{d^{j+1}}, j = 0, 1, 2....$
 2.2. Stop when $\|S_i^{j+1} - S_i^j\| < \epsilon$ and set $S_i = S_i^{j+1}$
3. Update the cluster assignments by relabelling each sample by $y = \underset{i}{arg\ max}\{S_i\}$.

Algorithm 1: F-LS-TWSVC clustering algorithm

6.8 Experimental Results

The TWSVC, LS-TWSVC and F-LS-TWSVC clustering methods were implemented by using MATLAB 8.1 running on a PC with Intel 3.40 GHz with 16 GB of RAM. The methods were evaluated on several benchmark datasets from UCI Machine Learning Repository ([29]).

6.8.1 Performance Measure for UCI Datasets

To compare the performance of the discussed clustering algorithm, we used the metric accuracy [25] as the performance criteria for UCI datasets. Given the kth cluster labels y_i where $i = 1, \ldots, m$, compute the corresponding similarity matrix $M \in R^{m\times m}$, where

$$M(i, j) = \begin{cases} 1 & : if \quad y_i = y_j \\ 0 & : \text{otherwise.} \end{cases} \tag{6.62}$$

Let, M_t is the similarity matrix computed by the true cluster label of the data set, and M_p corresponds to the label computed from the prediction of clustering method. Then, the metric accuracy of the clustering method is defined as the

$$Metric\ Accuracy = \frac{n_{00} + n_{11} - m}{m^2 - m} \times 100\,\%, \tag{6.63}$$

where n_{00} is the number of zeros in M_p and M_t, and n_{11} is the number of ones in M_p and M_t respectively.

6.8.2 Performance Measure for BSD

To establish the validity of our proposed formulations, we also perform experiments on the Berkeley Segmentation Dataset (BSD) [30] and for comparison we use *F-measure* [31] and Error Rate (ER) [32] as the performance criteria.

- *F-measure* can be calculated as

$$F - measure = \frac{2 \times Precision \times Recall}{Precision + Recall}, \tag{6.64}$$

with respect to human ground truth boundaries. Here,

$$Precision = \frac{TP}{TP + FP},$$

and

$$Recall = \frac{TP}{TP + FN}.$$

- ER can be calculated as

$$ER = \frac{FP + FN}{TT}, \tag{6.65}$$

where TP is number of true-detection object pixels, FP is the number of false-detection object pixels, FN is the number of false-detection not object pixels and TT is the total number of pixels present in the image.

For our simulations, we have considered RBF Kernel and the values of parameters like C, ν and sigma (kernel parameter) are optimized from the set of values $\{2^i | i = -9, -8, \ldots, 0\}$ using cross validation methodology [33]. The initial cluster labels and fuzzy membership values are optimized from FNNG initialization as discussed in Sect. 6.8.3.

6.8.3 Steps Involved in Initialization of Initial Fuzzy Membership Matrix via Fuzzy NNG

Traditionally, the initial labels of clustering can be generated randomly. However, in our algorithm discussed in Algorithm 1, we use fuzzy membership matrix as initial input. In [25], authors have shown via experiments that the results of plane based clustering methods strongly depends on the initial input of class labels. Hence taking motivation from initialization algorithm based on NNG [25, 34], we implement fuzzy NNG(FNNG) and provide output in the form of fuzzy membership matrix from FNNG method as the initial input to our algorithm. The main process of calculating FNNG is as follows:

1. For the given data set and a parameter p, construct p nearest neighbor undirected graph whose edges represents the distance between x_i (i=1, ..., m) and its p nearest neighbor.
2. From the graph, t clusters are obtained by associating the nearest samples. Further, construct a fuzzy membership matrix S_{ij} where $i = 1, \ldots m$ and $j = 1, \ldots t$ whose (i, j) entry can be calculated as follows,

$$S_{ij} = \frac{1}{d_{ij}}, \tag{6.66}$$

   where d_{ij} is the euclidean distance of the sample i with the jth cluster. If the current number of cluster t is equal to k, then stop. Else, go to step 3 or 4 accordingly.
3. If $t < k$, disconnect the two connected samples with the largest distance and go to step 2.
4. If $t > k$, compute the Hausdorff distance [35] between every two clusters among the t clusters and sort all pairs in ascending order. Merge the nearest pair of clusters into one, until k clusters are formulated, where the Hausdorff distance between two sets S_1 and S_2 of sample is defined as

$$h(S_1, S_2) = max\{\max_{i \in S_1}\{\min_{j \in S_2}||i - j||\}, \{\max_{i \in S_2}\{\min_{j \in S_1}||i - j||\}\}. \tag{6.67}$$

For the initialization of the CCCP in F-LS-TWSVC, i.e. for the value of an initial decision variable $[w_1^0 \quad b_1^0]$ we have implemented F-LS-TWSVM [36] classifier and obtained the solution for the aforementioned variables.

6.8.4 Computational Complexity

In [6], the authors have shown that TWSVM is approximately 4 times faster than SVM. The computational complexity of TWSVM is $(m^3/4)$, where m is the total size of training samples. In [12], the authors have shown that the solution of LS-TWSVM requires system of linear equations to be solved as opposed to the solution in TWSVM which requires system of linear equations along with two QPPs to be solved.

On the similar lines our algorithm F-LS-TWSVC essentially differs from TWSVC from the optimization problem involved i.e. in order to obtain k cluster plane parameters, we solve only two matrix inverse of $(n+1) \times (n+1)$ in linear case whereas TWSVC seeks to solve system of linear equations along with two QPPs. Table 6.1 shows the training time comparison among different algorithms with linear kernel on UCI dataset.

For nonlinear F-LS-TWSVC, solution requires inverse of the matrices with order $(m+1) \times (m+1)$ which can further be solved by (6.61) using SMW formula where we tend to solve inverse of two smaller dimension $(m_i \times m_i)$ and $((m-m_i) \times (m-m_i))$ matrices. Table 6.2 shows the training time comparison among different techniques with nonlinear kernel on UCI dataset.

Table 6.1 Average training time (in seconds) with linear kernel on UCI datasets

Data	TWSVC	LS-TWSVC	F-LS-TWSVC
Zoo	0.1262	0.0042	0.0052
Wine	0.0916	0.0033	0.0047
Iris	0.1645	0.0044	0.0051
Glass	0.2788	0.0062	0.0074
Dermatology	0.2666	0.0114	0.0160
Ecoli	0.2687	0.0115	0.0136
Compound	0.5570	0.0199	0.0225
Haberman	0.1156	0.0054	0.0068
Libas	0.4592	0.0319	0.0491
Page Blocks	7.6533	0.5316	0.8183
Optical Recognition	8.3640	0.1860	0.2220

Table 6.2 Average training time (in seconds) with nonlinear kernel on UCI datasets

Data	TWSVC	LS-TWSVC	F-LS-TWSVC
Zoo	1.1200	0.5300	0.7000
Wine	1.6272	0.8447	0.9677
Iris	1.0535	0.5314	0.6468
Glass	6.8200	2.1500	2.6000
Dermatology	12.1500	6.2700	6.9100
Ecoli	6.6280	2.9400	3.5111
Compound	17.6589	4.8600	5.3526
Haberman	3.1300	0.9593	1.1900
Libas	28.7700	19.0800	19.9400
Page Blocks	204.6000	64.5000	78.6512
Optical Recognition	420.5000	190.4100	210.3333

6.8.5 Experimental Results on UCI Datasets

In this section, we perform experiments on different UCI datasets with TWSVC, and compared its efficacy with proposed algorithms LS-TWSVC and F-LS-TWSVC respectively. The summary of UCI datasets is given in Table 6.3.

In [25], authors have reported clustering accuracy by considering whole dataset for learning the cluster hyperplanes. However, in our presentation of results, we have calculated training clustering accuracy as well as out of sample testing clustering accuracy along with reporting clustering accuracy on the whole dataset together. As a result, we have presented the results in two subsection discussed below.

6.8.5.1 Part 1 Results

Tables 6.4 and 6.5 summarizes the clustering accuracy results of our proposed F-LS-TWSVC and LS-TWSVC with TWSVC on several UCI benchmark datasets using linear and nonlinear kernel respectively. These tables show that metric accuracy of LS-TWSVC and TWSVC are comparable to each other, which further increases approximately $2-5\,\%$ on each datasets after incorporating fuzzy membership matrix. In Tables 6.4 and 6.5 we have taken results of kPC [22], PPC [23] and FCM [37] from [25].

Table 6.3 Summary of UCI datasets

Dataset	No. of instances	No. of features	No. of classes
Zoo	101	17	7
Wine	178	13	3
Iris	150	4	3
Glass	214	9	6
Dermatology	358	34	6
Ecoli	327	7	5
Compound	399	2	2
Haberman	306	3	2
Libras	360	90	15
Page blocks	5473	10	5
Optical recognition	5620	64	10

Table 6.4 Clustering accuracy with linear kernel on UCI datasets

Data	kPC [22]	PPC [23]	FCM [37]	TWSVC	LS-TWSVC	F-LS-TWSVC
Zoo	23.31	86.85	85.82	88.83	89.40	92.65
Wine	33.80	73.29	71.05	89.19	89.36	93.18
Iris	50.56	83.68	87.97	91.24	91.02	95.74
Glass	50.65	65.71	71.17	68.08	67.88	69.02
Dermatology	60.50	62.98	69.98	87.89	86.31	91.44
Ecoli	27.01	64.42	78.97	83.68	84.04	88.13
Compound	67.54	76.92	84.17	88.31	88.33	88.70
Haberman	60.95	60.95	49.86	62.21	62.14	62.21
Libras	49.90	81.37	51.89	89.42	89.64	90.14
Page blocks	–	–	–	79.88	79.58	81.01
Optical recognition	–	–	–	79.26	79.22	80.17

Table 6.5 Clustering accuracy with nonlinear kernel on UCI datasets

Data	kPC	PPC	TWSVC	LS-TWSVC	F-LS-TWSVC
Zoo	89.31	87.84	90.63	91.88	95.14
Wine	55.77	83.05	91.24	91.42	94.66
Iris	77.77	91.24	91.24	91.66	96.66
Glass	63.45	66.95	69.04	69.08	70.96
Dermatology	64.71	71.83	89.44	89.96	93.22
Ecoli	86.35	70.17	85.45	87.01	90.17
Compound	88.49	96.84	97.78	96.32	97.88
Haberman	61.57	61.57	61.26	62.14	62.74
Libras	85.32	87.79	90.08	90.56	92.01
Page blocks	–	–	80.78	80.42	82.38
Optical recognition	–	–	81.32	81.06	82.14

6.8.5.2 Part 2 Results

In this part, clustering accuracy was determined by following the standard 5-fold cross validation methodology [33]. Tables 6.6 and 6.7 summarizes testing clustering accuracy results of our proposed algorithms F-LS-TWSVC and LS-TWSVC with TWSVC on several UCI benchmark datasets.

Table 6.6 Testing clustering accuracy with linear kernel on UCI datasets

Data	TWSVC	LS-TWSVC	F-LS-TWSVC
Zoo	92.21 ± 3.23	93.56 ± 2.88	96.10 ± 2.18
Wine	85.88 ± 4.16	84.94 ± 4.89	90.92 ± 2.78
Iris	86.01 ± 8.15	86.57 ± 8.05	96.55 ± 1.23
Glass	65.27 ± 4.12	61.20 ± 5.26	65.41 ± 3.80
Dermatology	87.80 ± 2.39	88.08 ± 1.17	92.68 ± 2.42
Ecoli	80.96 ± 5.16	82.45 ± 4.96	86.23 ± 4.56
Compound	89.34 ± 3.53	90.70 ± 3.20	90.22 ± 3.29
Haberman	62.57 ± 4.06	60.63 ± 3.94	64.63 ± 3.94
Libas	87.31 ± 1.53	87.34 ± 0.64	88.52 ± 0.49
Page blocks	74.98 ± 4.07	74.63 ± 3.89	76.32 ± 3.12
Optical recognition	74.01 ± 4.78	73.33 ± 5.04	77.40 ± 4.32

Table 6.7 Testing clustering accuracy comparison with nonlinear kernel on UCI datasets

Data	TWSVC	LS-TWSVC	F-LS-TWSVC
Zoo	93.47 ± 3.96	94.76 ± 3.04	97.26 ± 2.68
Wine	87.66 ± 4.46	88.04 ± 4.98	92.56 ± 3.48
Iris	88.08 ± 7.45	89.77 ± 7.88	97.25 ± 2.23
Glass	67.27 ± 4.62	64.64 ± 5.66	68.04 ± 4.14
Dermatology	88.26 ± 3.49	88.77 ± 1.74	94.78 ± 2.90
Ecoli	83.28 ± 5.46	84.74 ± 5.07	88.96 ± 5.24
Compound	90.14 ± 3.68	90.98 ± 3.44	91.88 ± 3.55
Haberman	62.16 ± 4.26	60.03 ± 3.14	63.36 ± 3.44
Libas	88.16 ± 1.98	88.46 ± 1.06	90.05 ± 0.84
Page blocks	76.68 ± 5.22	75.99 ± 6.07	79.88 ± 5.51
Optical recognition	75.82 ± 5.78	75.32 ± 6.03	78.44 ± 4.11

6.8.6 Experimental Results on BSD Datasets

In this section we perform image segmentation on BSD dataset with proposed algorithm F-LS-TWSVC. Texture feature is one common feature used in image segmentation. Hence, we extract pixel-level texture feature from the images with the help of Gabor filter. Gabor filter [38] is a class of filters in which a filter of arbitrary orientation and scale is synthesized as linear combination of a set of "basis filter". It allows one to adaptively "steer "a filter to any orientation and scale, and to determine analytically the filter output as a function of orientation and scale. In our experiments, we use three level of scale (0.5, 1.0, 2.0) and four level of orientation

Table 6.8 F-measure and error rate on BSD dataset

Image-ID	F-measure		ER	
	TWSVC	F-LS-TWSVC	TWSVC	F-LS-TWSVC
3096	0.0250	0.0279	0.0538	0.0499
35070	0.0182	0.0427	0.2216	0.2001
42049	0.0215	0.0699	0.1249	0.0879
71046	0.0619	0.0625	0.2353	0.2280
86016	0.0491	0.0618	0.4806	0.3951
135069	0.0101	0.0141	0.0426	0.0380
198023	0.0500	0.0522	0.0742	0.0687
296059	0.0341	0.0369	0.0645	0.0616

Table 6.9 Training time (in seconds) comparison of image segmentation on BSD dataset

Image-ID	TWSVC	F-LS-TWSVC
3096	060.7049	020.6946
35070	168.8547	130.7123
42049	821.8520	510.9260
71046	118.2998	066.9686
86016	221.4578	130.2693
135069	395.4747	188.3213
198023	482.3645	200.9582
296059	275.1587	185.7195

$(0^o, 45^o, 90^o, 135^o)$. As a result, we have $12(3 \times 4)$ coefficients for each pixel of image. Finally, we use maximum (in absolute value) of the 12 coefficients for each pixels which represents the pixel-level wise Gabor features of an image. Further, this feature used as an input to FNNG which give us initial membership matrix for every pixels in different clusters. We have also use this Gabor filter to identify number of clusters present in the image.

Table 6.8 compare the performance of implemented F-LS-TWSVC with TWSVC methods on Barkeley Segmentation Dataset. It is noticeable that for better segmentation, the value of *F-measure* should be high and the value of ER should be less. Table 6.8 shows that the value of F-measure is high and the value of ER is less with F-LS-TWSVC than TWSVC (Table 6.9).

Figures 6.1, 6.2, 6.3 and 6.4 shows the segmentation results with F-LS-TWSVC and TWSVC respectively.

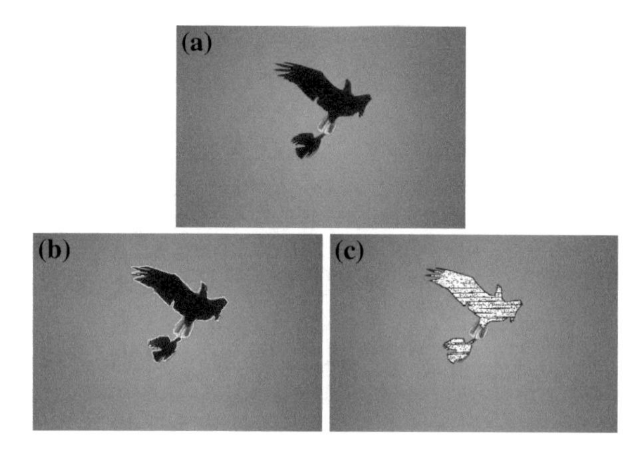

Fig. 6.1 Segmentation results **a** original image (ImageID-296059), **b** segmented image with F-LS-TWSVC and **c** segmented image with TWSVC

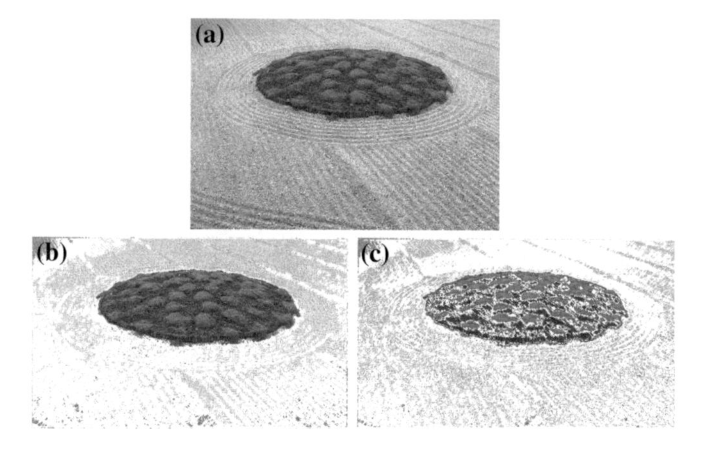

Fig. 6.2 Segmentation results **a** original image (ImageID-86016), **b** segmented image with F-LS-TWSVC and **c** segmented image with TWSVC

Fig. 6.3 Segmentation results **a** original image (ImageID-71046), **b** segmented image with F-LS-TWSVC and **c** segmented image with TWSVC

Fig. 6.4 Segmentation results **a** original image (ImageID-198023), **b** segmented image with F-LS-TWSVC, and **c** segmentation results with TWSVC

6.9 Conclusions

In this chapter, we review the variants of TWSVM for semi-supervised and unsupervised framework which are developed in the recent past. To begin with, this chapter discusses Laplacian SVM, Laplacian TWSVM, Laplacian LSTWSVM all variants of SVMs in semi-supervised settings. For unsupervised classification we have discussed K-Means, followed by plane based clustering algorithm which are on the lines of twin support vector machines and are termed as twin support vector clustering. We have also discussed our proposed work on fuzzy least squares twin support vector clustering and show its results on UCI as well as Image datasets.

References

1. Belkin, M., Niyogi, P., & Sindhwani, V. (2006). Manifold regularization: a geometric framework for learning from labeled and unlabeled examples. *The Journal of Machine Learning Research, 7*, 2399–2434.
2. Blum, A., & Mitchell, T. (1998). Combining labeled and unlabeled data with co-training. In *Proceedings of the Eleventh Annual Conference on Computational Learning Theory* (pp. 92–100).
3. Nigam, K., McCallum, A. K., Thrun, S., & Mitchell, T. (2000). Text classification from labeled and unlabeled documents using em. *Machine learning, 39*(2–3), 103–134.
4. Melacci, S., & Belkin, M. (2011). Laplacian support vector machines trained in the primal. *The Journal of Machine Learning Research, 12*, 1149–1184.
5. Zhu, X. (2008). *Semi-supervised learning literature survey, Computer Science TR (150)*. Madison: University of wisconsin.
6. Jayadeva, Khemchandani. R., & Chandra, S. (2007). Twin support vector machines for pattern classification. *IEEE Transactions on Pattern Analysis and Machine Intelligence, 29*(5), 905–910.
7. Suykens, J. A., & Vandewalle, J. (1999). Least squares support vector machine classifiers. *Neural processing letters, 9*(3), 293–300.
8. Kumar, M. A., & Gopal, M. (2009). Least squares twin support vector machines for pattern classification. *Expert Systems and its Applications, 36*(4), 7535–7543.
9. Zhang, Z., Zhen, L., Deng, N., & Tan, J. (2014). Sparse least square twin support vector machine with adaptive norm. *Applied Intelligence, 41*, 1097–1107.
10. Qi, Z., Tian, Y., & Shi, Y. (2012). Laplacian twin support vector machine for semi-supervised classification. *Neural Networks, 35*, 46–53.
11. Chen, W. J., Shao, Y. H., Deng, N. Y., & Feng, Z. L. (2014). Laplacian least squares twin support vector machine for semi-supervised classification. *Neurocomputing, 145*, 465–476.
12. Fung, G., & Mangasarian, O. L. (2001). Proximal support vector machine classifiers. In, F. Provost and R. Srikant (Eds.) *Proceedings of Seventh International Conference on Knowledge Discovery and Data Mining*, 77–86
13. Khemchandani, R., & Pal, A. (2016). Multicategory laplacian twin support vector machines. Applied Intelligence (To appear)
14. Anderberg, M. (1973). *Cluster Analysis for Applications*. New York: Academic Press.
15. Jain, A., Murty, M., & Flynn, P. (1999). Data clustering: a review. *ACM Computing Surveys (CSUR), 31*(3), 264–323.
16. Aldenderfer, M., & Blashfield, R. (1985). *Cluster Analysis*. Los Angeles: Sage.
17. QiMin, C., Qiao, G., Yongliang, W., & Xianghua, W. (2015). Text clustering using VSM with feature clusters. *Neural Computing and Applications, 26*(4), 995–1003.
18. Zhan, Y., Yin, J., & Liu, X. (2013). Nonlinear discriminant clustering based on spectral regularization. *Neural Computing and Applications, 22*(7–8), 1599–1608.
19. Tu, E., Cao, L., Yang, J., & Kasabov, N. (2014). A novel graph-based k-means for nonlinear manifold clustering and representative selection. *Neurocomputing, 143*, 109–122.
20. Liu, X., & Li, M. (2014). Integrated constraint based clustering algorithm for high dimensional data. *Neurocomputing, 142*, 478–485.
21. Bradley, P., & Mangasarian, O. (1997). Clustering via concave minimization. *Advances in Neural Information Processing Systems, 9*, 368–374.
22. Bradley, P., & Mangasarian, O. (2000). K-plane clustering. *Journal of Global Optimization, 16*(1), 23–32.
23. Shao, Y., Bai, L., Wang, Z., Hua, X., & Deng, N. (2013). Proximal plane clustering via eigenvalues. *Procedia Computer Science, 17*, 41–47.
24. Yang, Z., Guo, Y., Li, C., & Shao, Y. (2014). Local k-proximal plane clustering. *Neural Computing and Applications, 26*(1), 199–211.

25. Wang, Z., Shao, Y., Bai, L., & Deng, N. (2014). Twin support vector machine for clustering. *IEEE Transactions on Neural Networks and Learning Systems*,. doi:10.1109/TNNLS.2014.2379930.
26. Yuille, A. L., & Rangarajan, A. (2002). *The concave-convex procedure (CCCP)* (Vol. 2)., Advances in Neural Information Processing Systems Cambridge: MIT Press.
27. Khemchandani, R., & Pal, A. Fuzzy least squares twin support vector clustering. Neural Computing and its Applications (To appear)
28. Golub, G. H., & Van Loan, C. F. (1996). *Matrix computations* (3rd ed.). Baltimore: John Hopkins University Press.
29. Blake, C. L., & Merz, C. J. UCI Repository for Machine Learning Databases, Irvine, CA: University of California, Department of Information and Computer Sciences. http://www.ics.uci.edu/~mlearn/MLRepository.html.
30. Arbelaez, P., Fowlkes, C., & Martin, D. (2007). The Berkeley Segmentation Dataset and Benchmark. http://www.eecs.berkeley.edu/Research/Projects/CS/vision/bsds.
31. Mehrkanoon, S., Alzate, C., Mall, R., Langone, R., & Suykens, J. (2015). Multiclass semisupervised learning based upon kernel spectral clustering. *IEEE Transactions on Neural Networks and Learning Systems*, *26*(4), 720–733.
32. Wang, X. Y., Wang, T., & Bu, J. (2011). Color image segmentation using pixel wise support vector machine classification. *Pattern Recognition*, *44*(4), 777–787.
33. Duda, R., Hart, P., & Stork, D. (2001). *Pattern classification*. New York: Wiley.
34. Keller, J. M., Gray, M. R., & Givens, J. A. (1985). A fuzzy k-nearest neighbor algorithm. *IEEE Transactions on Systems, Man and Cybernetics*, *4*, 580–585.
35. Hausdorff, F. (1927). *Mengenlehre*. Berlin: Walter de Gruyter.
36. Sartakhti, J. S., Ghadiri, N., & Afrabandpey, H. (2015). Fuzzy least squares twin support vector machines. arXiv preprint arXiv:1505.05451.
37. Wang, X., Wang, Y., & Wang, L. (2004). Improving fuzzy c-means clustering based on feature-weight learning. *Pattern Recognition Letters*, *25*, 1123–1132.
38. Manjunath, B. S., & Ma, W. Y. (1996). Texture features for browsing and retrieval of image data. *IEEE Transactions on Pattern Analysis and Machine Intelligence*, *18*(8), 837–842.

Chapter 7
Some Additional Topics

7.1 Introduction

This chapter is devoted to the study of certain additional topics on twin support vector machines for classification. Specifically, these topics are kernel optimization in TWSVM, knowledge based TWSVM and a recently introduced formulation of Twin Support Tensor Machine (TWSTM) for matrix data classification. We are including these topics for our discussion here because of their novelty and potential for real life applications.

This chapter consists of five main sections namely, Optimal Kernel Selection in Twin Support Vector Machines, Knowledge Based Twin Support Vector Machines and Variants, Support Tensor Machines: A Brief Review, and Twin Support Tensor Machines.

Our presentation here is based on Khemchandani et al. [1, 2], Kumar et al. [3], Cai et al. [4], Zhao et al. [5] and Gao et al. [6].

7.2 Optimal Kernel Selection in Twin Support Vector Machines

It is well known that kernel based methods have proven to be a powerful tool for solving classification and regression problems. Generally, kernels are chosen by predefining a kernel function (Gaussian, polynomial etc.) and then adjusting the kernel parameters by means of a tuning procedure. The classifier's/regressor's performance on a subset of the training data, commonly referred to as the validation set, is usually the main criterion for choosing a kernel. But this kernel selection procedure is mostly adhoc and can be computationally expensive.

In recent years, several authors have proposed the use of a kernel that is obtained as an 'optimal' non-negative linear combination of finitely many 'basic' kernels. The kernel so chosen is termed as an 'optimal kernel'. This optimal kernel is

Jayadeva et al., *Twin Support Vector Machines*, Studies in Computational Intelligence 659, DOI 10.1007/978-3-319-46186-1_7

constructed to enhance the generalization performance of the underlying classifier/regressor.

For determining an 'optimal kernel' in the framework of SVM learning, Lanckriet et al. [7] employed semi-definite programming in a very effective manner and showed that this 'optimal' kernel performs better than the individual kernels of the chosen kernel family. One interesting feature of this formulation is its sound theoretical foundation because the class of semi-definite programming problems is a very well studied class in terms of optimality conditions, duality and algorithms. Xiong et al. [8] also studied this type of problem and showed that these optimal non-negative coefficients can be learnt from the data by employing a gradient based procedure. Later Jayadeva et al. [9] studied the same problem as that of Xiong et al. [8], but instead of a gradient based approach, they showed that these coefficients can be learnt from the data by simply solving a generalized eigenvalue problem.

In the context of Kernel Optimization for Fisher Discriminant Analysis, Fung et al. [10] developed an iterative method that alternates between optimizing the weight vector and the Gram matrix. This method is based on a quadratic programming formulation of the Fisher discriminant analysis given by Mika et al. [11].

In this section, we address the problem of optimal kernel selection in the context of TWSVM, and taking motivation from Fung et al. [10], present an algorithm in the spirit of iterative alternating optimization proposed by Bezdek and Hathaway [12]. Since TWSVM determines a pair of hyperplanes, the problem of optimal kernel selection for TWSVM involves selecting an optimal pair of kernels which are obtained by solving a pair of TWSVM - type problems and pair of convex optimization problems. In each pair, the first set of problems finds the optimal classifier, while the second finds the optimal kernel matrix. Thus, the proposed method alternates between optimizing the vector corresponding to the classifier and determining a set of non-negative weights, that define the optimal kernel matrix in terms of a weighted linear combination of elements from a given family of primary kernels. The task of finding the optimal kernel is incorporated in the optimization problem. The whole algorithm is in the spirit of an alternating optimization algorithm, and therefore its convergence is assured by the strict convexity of the constituent quadratic programming problems. Experimental results illustrate the efficacy of the proposed algorithm.

An added advantage of the algorithm is the provision of finding the optimal pair of combinations of kernels for different classes, which is not present in SVM scenario. In particular it may happen that the Gaussian kernel with hyper-parameter σ_1 is significant for class 1, whereas the Gaussian kernel with hyper-parameter σ_2 is significant for class 2. The usual kernel SVM will be forced to take the same σ rather than σ_1 for class 1 and σ_2 for class 2. In our formulation of the kernel optimization for TWSVM, this provision is inbuilt. Therefore, from the optimal values of weights, we shall possibly be able to better understand the structure of dataset. Such an interpretation about the dataset can possibly not be drawn in the SVM framework.

Our presentation here is based on Khemchandani et al. [1].

7.2.1 Problem Formulation and an Alternating Optimization Algorithm

Let us recall our discussion on kernel TWSVM formulation from Chap. 3. Let the patterns to be classified be denoted by a set of m row vectors X_i, $(i = 1, 2, \ldots, m)$ in the n-dimensional real space $\mathbb{R}^n$, and let $y_i \in \{1, -1\}$ denote the class to which the ith pattern belongs. Matrices A and B represent data points belonging to classes $+1$ and -1, respectively. Let the number of patterns in classes $+1$ and -1 be given by m_1 and m_2, respectively. Therefore, the matrices A and B are of sizes $(m_1 \times n)$ and $(m_2 \times n)$, respectively.

In order to obtain the nonlinear classifiers, we consider the following kernel generated surfaces

$$K(x^T, C^T)u_1 + b_1 = 0, \quad \text{and} \quad K(x^T, C^T)u_2 + b_2 = 0, \tag{7.1}$$

where $C^T = [A \quad B]^T$, $(u_i, b_i) \in (\mathbb{R}^m \times \mathbb{R})(i = 1, 2)$ and K is an appropriately chosen kernel. The nonlinear classifiers are obtained by solving the following optimization problem (KTWSVM1) and its dual (KDTWSVM1)

$$(KTWSVM1) \qquad \underset{(u_1, b_1, q_1)}{\text{Min}} \quad \frac{1}{2}||(K(A, C^T)u_1 + e_1 b_1)||^2 + c_1 e_2^T q_1$$
$$\text{subject to}$$
$$-(K(B, C^T)u_1 + e_2 b_1) + q_1 \geq e_2,$$
$$q_1 \geq 0, \tag{7.2}$$

and

$$(KDTWSVM1) \qquad \underset{\alpha}{\text{Max}} \quad e_2^T\alpha - \frac{1}{2}\alpha^T R(S^T S)^{-1} R^T \alpha$$
$$\text{subject to}$$
$$0 \leq \alpha \leq c_1, \tag{7.3}$$

respectively. Here $c_1 > 0$ is a trade-off parameter, α is the vector of Lagrange multipliers, and the matrices S and R are defined as

$$S = \begin{bmatrix} K(A, C^T) & e_1 \end{bmatrix}, \; R = \begin{bmatrix} K(B, C^T) & e_2 \end{bmatrix}; \tag{7.4}$$

with, e_1 and e_2 as vectors of ones of appropriate dimension.

The augmented vector $z^{(1)} = [u_1, b_1]^T$ is determined by

$$z^{(1)} = -(S^T S)^{-1} R^T \alpha, \tag{7.5}$$

which gives the plane $K(x^T, C^T)u_1 + b_1 = 0$.

Furthermore, by reversing the roles of $K(A, C^T)$ and $K(B, C^T)$ in (7.2), we obtain the optimization problem (KTWSVM2) and its dual (KDTWSVM2) for the plane $K(x^T, C^T)u_2 + b_2 = 0$ as

$$(KTWSVM2) \qquad \underset{(u_2,b_2,q_2)}{\text{Min}} \quad \frac{1}{2}||(K(B, C^T)u_2 + e_2 b_2)||^2 + c_2 e_1^T q_2$$
$$\text{subject to}$$
$$(K(A, C^T)u_2 + e_1 b_2) + q_2 \geq e_1,$$
$$q_2 \geq 0, \tag{7.6}$$

where $c_2 > 0$ is a parameter; and

$$(KDTWSVM2) \qquad \underset{\gamma}{\text{Max}} \quad e_1^T \gamma - \frac{1}{2}\gamma^T L(N^T N)^{-1} L^T \gamma$$
$$\text{subject to}$$
$$0 \leq \gamma \leq c_2. \tag{7.7}$$

Here, the matrices L and N are defined as

$$L = [K(A, C^T) \quad e_1]; \quad N = [K(B, C^T) \quad e_2]. \tag{7.8}$$

The augmented vector $z^{(2)} = [u_2, b_2]^T$ is determined by

$$z^{(2)} = (N^T N)^{-1} L^T \gamma, \tag{7.9}$$

which gives the plane $K(x^T, C^T)u_2 + b_2 = 0$.

A new data sample $x \in \mathbb{R}^n$ is then assigned to class $r(r = 1, 2)$, depending on which of the two planes given by (7.1) it lies closest to. Thus

$$\text{class}(x) = arg \underset{(r=1,2)}{\text{Min}} (d_r(x)) \tag{7.10}$$

$$\text{where} \quad d_r(x) = \left(\frac{|K(x, C)u_r + b_r|}{||u_r||} \right), \tag{7.11}$$

and $||u||$ is the L_2 norm of vector u.

As explained in Chap. 3, we add regularization terms to the expressions $(S^T S)$ and $(N^T N)$ in (7.3) and (7.7), so as to assure strict convexity of the corresponding quadratic programming problems. In practice, if the number of patterns in classes 1 or -1 is large, then the rectangular kernel technique proposed by Fung and Mangasarian

[13] can be applied to reduce the dimensionality of the kernel matrix, i.e. we can replace $K(A, C^T)$ by a rectangular kernel of the type $K(A, Q^T)_{m_1 \times t}$, where Q is a $t \times n$ random sub matrix of C, with $t << m$. In fact, t may be as small as 0.01 m. Thus, the size of kernel matrix becomes $m_1 \times t$, thereby requiring the inversion of a matrix of order $(t+1) \times (t+1)$ only.

The optimization problems (KDTWSVM1) and (KDTWSVM2) can be rewritten as

$$(KDTWSVM1) \qquad \underset{\alpha}{\text{Max}} \quad e_2^T \alpha - \frac{1}{2}\alpha^T T^{(1)} \alpha$$
$$\text{subject to}$$
$$0 \leq \alpha \leq c_1,$$

and

$$(KDTWSVM2) \qquad \underset{\gamma}{\text{Max}} \quad e_1^T \gamma - \frac{1}{2}\gamma^T T^{(2)} \gamma$$
$$\text{subject to}$$
$$0 \leq \gamma \leq c_2,$$

where the matrices $T^{(1)}$ and $T^{(2)}$ are defined by $T^{(1)} = R(S^T S + \epsilon I)^{-1} R^T$ and $T^{(2)} = L(N^T N + \epsilon I)^{-1} L^T$, respectively. Here the matrices R and S; and, L and N are as defined in (7.4), and (7.8) respectively. Given a kernel such as a Gaussian, or a polynomial, we compute $T^{(1)}$ and $T^{(2)}$ and solve (KDTWSVM1) and (KDTWSVM2), respectively, to obtain the kernel generated surfaces (7.1).

Let us now suppose that instead of the pair of kernels $T^{(1)}$ and $T^{(2)}$ being defined by a single kernel function such as a Gaussian or a polynomial, an optimal pair of kernels $\mathbf{J}^{(1)}$ and $\mathbf{J}^{(2)}$ corresponding to (KDTWSVM1) and (KDTWSVM2), respectively, are chosen as follows

$$\mathbf{J}^{(1)} = \sum_{j=1}^{p} \mu_j^{(1)} T_j^{(1)}; \quad \mathbf{J}^{(2)} = \sum_{j=1}^{p} \mu_j^{(2)} T_j^{(2)},$$

where $\mu_j^1, \mu_j^2 \geq 0$. Here, similar to $T^{(1)}$ and $T^{(2)}$, $T_j^{(1)}$ (and $T_j^{(2)}$) are constructed from the basic 'p' kernel matrices K_j, $(j = 1, 2, \ldots, p)$.

As it is pointed out in Lanckriet et al. [7], the set of basic kernels can be seen as a predefined set of initial guess of the kernel matrices. We also note that the set of basic kernels could contain very different kernel matrix models, e.g., linear, Gaussian, polynomial, or kernels with different hyper parameters.

Taking motivation from Fung et al. [10], we present an iterative alternating algorithm, termed as (A-TWSVM), for the determination of the optimal pair of kernel matrices $\mathbf{J}^{(1)}$ and $\mathbf{J}^{(2)}$. This method alternates between optimizing the decision vector (α, γ), and the scalars $\mu_j^{(1)}, \mu_j^{(2)} \geq 0, j = 1, 2, \ldots, p$, corresponding to a pair of optimal kernel matrices $\mathbf{J}^{(1)} = \sum_{j=1}^{p} \mu_j^{(1)} T_j^{(1)}$, and $\mathbf{J}^{(2)} = \sum_{j=1}^{p} \mu_j^{(2)} T_j^{(2)}$.

Instead of fine tuning of the kernel parameters for a predetermined kernel via cross-validation, we can optimize the set of values $\mu_j^{(1)}, \mu_j^{(2)} \geq 0, j = 1, 2, \ldots, p$, in order to obtain a positive semi-definite linear combination $\mathbf{J}^{(1)} = \sum_{j=1}^{p} \mu_j^{(1)} T_j^{(1)}$, and $\mathbf{J}^{(2)} = \sum_{j=1}^{p} \mu_j^{(2)} T_j^{(2)}$.

The formulation presented here involves solving two pairs of optimization problems iteratively. The first pair i.e. (7.12)–(7.13), deals with the optimization of α and γ for fixed μ_1 and μ_2.

$$(ATWSVM1) \qquad \underset{\alpha}{\text{Max}} \quad e_2^T \alpha - \frac{1}{2} \alpha^T \left(\sum_{j=1}^{p} \mu_j^{(1)} T_j^{(1)} \right) \alpha$$
$$\text{subject to}$$
$$0 \leq \alpha \leq c_1, \tag{7.12}$$

and

$$(ATWSVM2) \qquad \underset{\gamma}{\text{Max}} \quad e_1^T \gamma - \frac{1}{2} \gamma^T \left(\sum_{j=1}^{p} \mu_j^{(2)} T_j^{(2)} \right) \gamma$$
$$\text{subject to}$$
$$0 \leq \gamma \leq c_2. \tag{7.13}$$

The initial values of $\mu_j^{(i)}, i = 1, 2, j = 1, 2, \ldots, p$ are set to one. Once the values of α and γ are known, $z^{(1)}$ and $z^{(2)}$ are obtained from (7.5) and (7.9), and the classifier (7.1) is determined. With the values of α and γ as obtained from (7.12) and (7.13) respectively, the second part of the algorithm then finds the optimal weights $\mu_j^{(1)}, \mu_j^{(2)} \geq 0$, $j = 1, 2, \ldots, p$ by solving the pair of optimization problems (7.14)–(7.15)

$$\text{(ATWSVM3)} \qquad \underset{\substack{e^T \mu_1 = p, \\ \mu_1 \geq 0}}{\text{Min}} \ \frac{1}{2} {\mu_1}^T \mu_1 + \frac{1}{2} \sum_{j=1}^{p} \mu_j^{(1)} \alpha^T T_j^{(1)} \alpha, \tag{7.14}$$

and

$$\text{(ATWSVM4)} \qquad \underset{\substack{e^T\mu_2=p,\\ \mu_2\geq 0}}{\text{Min}} \quad \frac{1}{2}{\mu_2}^T\mu_2 + \frac{1}{2}\sum_{i=j}^{p}\mu_j^{(2)}\gamma^T T_j^{(2)}\gamma, \tag{7.15}$$

where e is a vector of ones of appropriate dimension. With the updated values of $\mu_j^{(i)}$, $(i = 1, 2)$; $(j = 1, 2 \ldots, p)$, (ATWSVM1)-(ATWSVM2) are solved to obtain the updated classifier. We continue solving these two pairs of optimization problems alternatively and stop when the prescribed termination criteria is met. The termination criteria for A-TWSVM is taken to be the same as that of A-KFD (Fung et al. [10]). Thus, we stop when the algorithm runs for a predefined maximum number of iterations, or when the change in the training accuracy evaluated with updated pair of kernel matrices $\sum_{j=1}^{p}\mu_j^{(1)}T_j^{(1)}$ and $\sum_{j=1}^{p}\mu_j^{(2)}T_j^{(2)}$ in successive iterations is within some tolerance $\epsilon > 0$ (the tolerance is user defined and we have taken $\epsilon = 10^{-2}$ in our numerical implementation).

In (ATWSVM3) and (ATWSVM4) the constraints $e^T\mu_1 = p$ and $e^T\mu_2 = p$ are significant. This is because if we normalize $T_j^{(i)}$; $(i = 1, 2)$; $(j = 1, 2, \ldots, p)$ so that all the kernel matrices have the same trace m, then the constraints $e^T\mu_1 = p$ and $e^T\mu_2 = p$ guarantee that the trace of $\mathbf{J}^{(i)}$, $(i = 1, 2)$ is also constant, the constant being equal to mp. Therefore $\mathbf{J}^{(i)}$, $(i = 1, 2)$ will belong to a family of positive semidefinite matrices with constant trace. In view of Lanckriet et al. [7], this condition ensures that there is no over fitting of data. In this context it may be remarked that no such constraints on μ's are enforced in A-KFD (Fung et al. [10]).

In this context, we may note that the kernel matrices corresponding to large $\mu_j^{(i)}$, $(i = 1, 2)$; $(j = 1, 2 \ldots, p)$ are important in describing the data, and are significant in characterizing the properties and structure of the data, whereas those with small $\mu_j^{(i)}$, $(i = 1, 2)$; $(j = 1, 2 \ldots, p)$ are less significant. Here, we observe that two different combination of kernels may be significant for two classes. The optimal values of $\mu_j^{(i)}$, $(i = 1, 2)$, $(j = 1, 2, \ldots, p)$ will determine the significance of the corresponding kernel for determining the structure of the data of that class.

A new data sample $x \in \mathbb{R}^n$ is assigned to class r $(r = 1, 2)$, depending on which of the two planes given by (7.1) it lies closest to. Thus

$$\text{class}(x) = arg \underset{(r=1,2)}{\text{Min}} (d_r(x))$$

$$\text{where} \quad d_r(x) = \left(\frac{\left|\sum_{j=1}^{p}\mu_j^{(r)}K_j(x, C)u^{(r)} + b_r\right|}{||u^{(r)}||}\right),$$

and $||u||$ is the L_2-norm of vector u.

In view of the details of the algorithm as discussed above, we can interpret A-TWSVM as an alternating optimization problem (AO) (Bezdek and Hathaway [12]). Since each of the two optimization problems in both pairs of A-TWSVM are strictly convex, they have unique optimal solutions. Therefore, the convergence of the proposed method is assured by employing the basic convergence theorem of Bezdek and Hathaway [12]. This theorem states that if the objective function is a strictly convex function on its domain that has a minimizer at which Hessian of the function is continuous and positive definite, then AO will converge q-linearly to the minimizer using any initialization.

7.2.2 Experimental Results

The performance of A-TWSVM was assessed by using the pair of kernel matrices $T_j = (T_j^{(1)}, T_j^{(2)})$ $(j = 1, 2, 3)$, and with the optimal pair of kernel matrices $(\mathbf{J}^{(1)}, \mathbf{J}^{(2)})$. The kernel matrices T_1, T_2 and T_3 are respectively defined by the linear kernel K_1, the Gaussian kernel K_2 and the polynomial kernel of degree two, namely, K_3. In the experiments, we have considered two sets of combinations of basic kernels. In the first case the pair of kernels is obtained as a non-negative linear combination of kernel matrices T_1, T_2 and T_3, whereas in the second case the pair of kernels is obtained as a non-negative linear combination of Gaussian kernels only, but with different variances (σ). Thus, for the first case the pair is $(\mathbf{J}_{123}^{(1)}, \mathbf{J}_{123}^{(2)})$ whereas for the second case the pair is $(\mathbf{J}_2^{(1)}, \mathbf{J}_2^{(2)})$. Here the suffices indicate that in the first case all the three kernels are used whereas for the second case only Gaussian kernels are used.

While implementing an individual kernel the optimal value of the kernel parameter is obtained by following the standard tuning procedure i.e. we have taken 10% of the dataset for finding the optimal value of the parameter. Further, while implementing A-TWSVM for finding the optimal pair of Gaussian kernels i.e. $(\mathbf{J}_2^{(1)}, \mathbf{J}_2^{(2)})$ we have chosen variances of the two Gaussian kernels in the neighborhood of $\sigma = \sigma_1$. Here, σ_1 has been obtained by tuning while implementing TWSVM with the individual kernel K_2. Thus, variance of one Gaussian is σ_1 and variances of other two Gaussian's are chosen randomly from an interval around σ_1. Similarly, while implementing A-TWSVM for $(\mathbf{J}_{123}^{(1)}, \mathbf{J}_{123}^{(2)})$ variance for K_2 is σ_1. For all the examples, the interval $[0.1, 2]$ suffices. Further, while implementing (ATWSVM3)-(ATWSVM4), the kernel matrices are normalized using

$$T_{st}^{(i)} = \frac{T^{(i)}(x_s, x_t)}{\sqrt{T^{(i)}(x_s, x_t)T^{(i)}(x_s, x_t)}}, \quad (i = 1, 2)(s, t = 1, 2, \ldots, m). \tag{7.16}$$

to ensure that the kernels $T_j^{(1)}$ and $T_j^{(2)}$ $(j = 1, 2, 3)$ have trace equal to m. This inturn ensures the validity of the constraints $e^T\mu_1 = 3$ and $e^T\mu_2 = 3$.

The optimal values of $\mu_j^{(1)}, \mu_j^{(2)} \geq 0, j = 1, 2, \ldots, p$ in A-TWSVM, are obtained by solving the problem iteratively. Using these optimal values of $\mu_j^{(1)}, \mu_j^{(2)} \geq 0$, $j = 1, 2, \ldots, p$, a pair of hyperplanes are built and then the performance on the testing set is evaluated. In our case $p = 3$.

The proposed method was evaluated on five datasets namely Ionosphere, Heart statlog, German Credit, Hepatitis and Sonar chosen from the UCI Machine Learning

Repository (Blake and Merz [14]), and the test set accuracy was determined by following the standard ten-fold cross-validation methodology (Duda et al. [15]). The hardware consisted of a PC with an Intel P4 processor (3 GHz) and 1 GB RAM.

First dataset which we have considered is Ionosphere dataset. This data classifies the radar returns as "good" or "bad" signifying if there is or isn't any evidence of structure in the ionosphere is observed. There are a total of 351 patterns, 224 are "good", and 127 are "bad" patterns and each having 34 continuous attributes. The Heart-statlog dataset has 270 instances, 150 patterns don't have heart disease whereas in 120 presence of heart disease is predicted. This dataset has 13 continuous attributes. The German credit data set is taken from Institute für Statistik und Okonometrie Hamburg University. It has 13 attributes selected from the original 20 attributes. There are a total of 1000 instances of which 700 are "good", and 300 are "bad". The Hepatitis data set is a binary data set from the hepatitis domain with each instance having 19 attributes like sex, age, steroid, antiviral, fatigue, bilirubin, albumin etc. that determine the disease. There are 155 instance of which 32 are "die" and 123 are "alive". Fifth dataset is Sonar dataset. This is a two class identification problem of underground targets ("rock", or "mines") on the basis of 60 attributes. There are a total of 208 patterns (111 "mines", and 97 "rocks" patterns).

For the implementation of A-TWSVM we have fixed the maximum number of iterations to be 10. Thus, the algorithm terminates after 10 iterations, or if there is a small change (as per our tolerance $\epsilon = 10^{-2}$) in the test set accuracy. The tables indicate (mean $\pm$ standard deviation) performance for each classifier on each dataset. Tables 7.1, 7.2, 7.3, 7.4 and 7.5 summarize the performance of the optimal kernel selection approach for twin support vector machines on some benchmark datasets available at the UCI machine learning repository. In practice we found that the our algorithm converges in 3–4 iterations on all the datasets. The average run time (in seconds) is also reported in the tables along with the average number of iterations. In each table μ_1 and μ_2 are the weights in the optimal non-negative linear combination of kernels in $\mathbf{J}_{123}^{(1)}$ and $\mathbf{J}_{123}^{(2)}$; and $\mathbf{J}_2^{(1)}$ and $\mathbf{J}_2^{(2)}$.

From the tables, we observe that sum of $\mu_j^{(i)}$, $(i = 1, 2)$; $(j = 1, 2, 3)$ is three and the kernels corresponding to the large $\mu_j^{(i)}$, $(i = 1, 2)$; $(j = 1, 2, 3)$ are important than the one with smaller values. This set of information may play an important role while learning the structure of the data. Further, we conclude that for a given classification problem the optimally combined kernel generally yields better test set accuracy than individual kernels.

From Table 7.1, we conclude that combination of three Gaussian's improves the test set accuracy for Ionosphere dataset as compare to linear, Gaussian and polynomial kernel. The same conclusion can be drawn for Heart-statlog, German Credit, Hepatitis and Sonar dataset. Further, from the tables we observe that when the two classes have different structures then the two different kernels are used to predict the structure of the data. This set of information can be obtained from the values of $\mu_j^{(i)}$, $(i = 1, 2)$; $(j = 1, 2, 3)$. For example, in Table 7.4, when the three different kernels are combined for Hepatitis dataset, for class 1 polynomial kernel of degree 2 dominates the linear and the Gaussian, while for class –1, the Gaussian dominates

Table 7.1 Percentage test set accuracy on the ionosphere dataset

Kernel	$(T_1^{(1)}, T_1^{(2)})$	$(T_2^{(1)}, T_2^{(2)})$	$(T_3^{(1)}, T_3^{(2)})$	$(\mathbf{J}_{123}^{(1)}, \mathbf{J}_{123}^{(2)})^*$	$(\mathbf{J}_2^{(1)}, \mathbf{J}_2^{(2)})^{**}$
TSA	89.17 ± 4.58	93.14 ± 4.28	89.17 ± 5.25	93.15 ± 4.82	94.29 ± 5.11
μ_1	–	–	–	(0, 3, 0)	(0.57, 1.00, 1.43)
μ_2	–	–	–	(0, 3, 0)	(0.41, 0.85, 1.74)
Iterations	–	–	–	2 ± 0	2.5 ± 1.75
Time (s)	1.16	1.20	1.51	3.22	3.85

$^*(\mathbf{J}_{123}^{(1)}, \mathbf{J}_{123}^{(2)})$ are obtained as non-negative linear combination of kernel matrices T_1, T_2 and T_3.
$^{**}(\mathbf{J}_2^{(1)}, \mathbf{J}_2^{(2)})$ are obtained as non-negative linear combination of Gaussian with different hyper parameters

Table 7.2 Percentage test set accuracy on the heartstatlog dataset

Kernel	$(T_1^{(1)}, T_1^{(2)})$	$(T_2^{(1)}, T_2^{(2)})$	$(T_3^{(1)}, T_3^{(2)})$	$(\mathbf{J}_{123}^{(1)}, \mathbf{J}_{123}^{(2)})^*$	$(\mathbf{J}_2^{(1)}, \mathbf{J}_2^{(2)})^{**}$
TSA	82.96 ± 4.44	81.85 ± 5.35	81.11 ± 6.07	83.47 ± 4.61	83.70 ± 6.46
μ_1	–	–	–	(0, 2.58, 0.42)	(1.03, 1.01, 0.96)
μ_2	–	–	–	(0, 3, 0)	(1.04, 1.02, 0.94)
Iterations	–	–	–	2.4 ± 0.4	2.1 ± 0.54
Time (s)	0.71	0.97	0.80	2.41	2.31

$^*(\mathbf{J}_{123}^{(1)}, \mathbf{J}_{123}^{(2)})$ are obtained as non-negative linear combination of kernel matrices T_1, T_2 and T_3.
$^{**}(\mathbf{J}_2^{(1)}, \mathbf{J}_2^{(2)})$ are obtained as non-negative linear combination of Gaussian with different hyper parameters

Table 7.3 Percentage test set accuracy on the German credit dataset

Kernel	$(T_1^{(1)}, T_1^{(2)})$	$(T_2^{(1)}, T_2^{(2)})$	$(T_3^{(1)}, T_3^{(2)})$	$(\mathbf{J}_{123}^{(1)}, \mathbf{J}_{123}^{(2)})^*$	$(\mathbf{J}_2^{(1)}, \mathbf{J}_2^{(2)})^{**}$
TSA	70.40 ± 3.01	70.90 ± 4.95	70.20 ± 3.43	70.90 ± 4.95	71.80 ± 2.93
μ_1	–	–	–	(0, 3, 0)	(0, 3, 0)
μ_2	–	–	–	(0, 3, 0)	(0, 3, 0)
Iterations	–	–	–	1.9 ± 0.3	2.00 ± 0.0
Time (s)	9.85	14.04	12.88	51.63	51.20

$^*(\mathbf{J}_{123}^{(1)}, \mathbf{J}_{123}^{(2)})$ are obtained as non-negative linear combination of kernel matrices T_1, T_2 and T_3.
$^{**}(\mathbf{J}_2^{(1)}, \mathbf{J}_2^{(2)})$ are obtained as non-negative linear combination of Gaussian with different hyper parameters

the other two kernels. Similar behavior is also observed when Gaussian with different hyper-parameters are used for training the Hepatitis and Sonar datasets.

Table 7.4 Percentage test set accuracy on the hepatitis dataset

Kernel	$(T_1^{(1)}, T_1^{(2)})$	$(T_2^{(1)}, T_2^{(2)})$	$(T_3^{(1)}, T_3^{(2)})$	$(\mathbf{J}_{123}^{(1)}, \mathbf{J}_{123}^{(2)})^*$	$(\mathbf{J}_2^{(1)}, \mathbf{J}_2^{(2)})^{**}$
TSA	79.83 ± 8.27	79.17 ± 9.00	79.83 ± 7.31	79.83 ± 8.85	82.46 ± 8.89
μ_1	–	–	–	(0, 0.33, 2.67)	(0.32, 0.17, 2.51)
μ_2	–	–	–	(0.67, 1.46, 0.87)	(1, 1.01, 0.99)
Iterations	–	–	–	2 ± 1.1	3.7 ± 0.9
Time (s)	0.30	0.28	0.35	0.95	1.89

$^*(\mathbf{J}_{123}^{(1)}, \mathbf{J}_{123}^{(2)})$ are obtained as non-negative linear combination of kernel matrices T_1, T_2 and T_3.
$^{**}(\mathbf{J}_2^{(1)}, \mathbf{J}_2^{(2)})$ are obtained as non-negative linear combination of Gaussian with different hyper parameters

Table 7.5 Percentage test set accuracy on the sonar dataset

Kernel	$(T_1^{(1)}, T_1^{(2)})$	$(T_2^{(1)}, T_2^{(2)})$	$(T_3^{(1)}, T_3^{(2)})$	$(\mathbf{J}_{123}^{(1)}, \mathbf{J}_{123}^{(2)})^*$	$(\mathbf{J}_2^{(1)}, \mathbf{J}_2^{(2)})^{**}$
TSA	65.49 ± 5.80	80.34 ± 7.31	72.23 ± 4.35	85.17 ± 6.61	85.88 ± 3.34
μ_1	–	–	–	(0, 2.67, 0.33)	(0, 3, 0)
μ_2	–	–	–	(0.7, 2.89, 0.4)	(0, 0, 3)
Iterations	–	–	–	2 ± 0	2 ± 0.5
Time (s)	1.17	2.47	2.32	5.60	6.24

$^*(\mathbf{J}_{123}^{(1)}, \mathbf{J}_{123}^{(2)})$ are obtained as non-negative linear combination of kernel matrices T_1, T_2 and T_3.
$^{**}(\mathbf{J}_2^{(1)}, \mathbf{J}_2^{(2)})$ are obtained as non-negative linear combination of Gaussian with different hyper parameters

7.3 Knowledge Based Twin Support Vector Machines and Variants

Classification with prior knowledge has now become an important field of research, e.g. classification of medical datasets where only expert's knowledge is prescribed and no training data is available. In an earlier work of Fung et al. [16], prior knowledge in the form of multiple polyhedral sets, each belonging to one of the two classes has been introduced into reformulation of linear support vector machine classifier. Fung et al. [16] have discussed 1-norm formulation of knowledge based classifier which they termed as Knowledge Based SVM (KBSVM) classifier. Here, with the use of theorems of alternative (Mangasarian [17]), polyhedral knowledge set is reformulated into a set of inequalities, with which prior data can be embedded into the linear programming formulation. Working on the lines of Fung et al. [16], Khemchandani et al. [18] have proposed an extremely simple and fast (KBPSVM) classifier in the light of PSVM formulation. They have used 2-norm approximation of knowledge sets and slack variables in their formulation instead of 1-norm used in KBSVM. Experimental results presented in [18] show that KBPSVM performs better than KBSVM in terms of both generalization and training speed.

Given this background, an interesting and motivating problem is: incorporation of polyhedral knowledge sets into SVM classifier based on two non-parallel hyperplanes. In this section we have addressed this problem by incorporating polyhedral knowledge sets into TWSVM/LSTWSVM formulations. For TWSVM and LSTWSVM formulations we shall refer to Chaps. 3 and 5 respectively. This new formulation, termed as (KBTSVM)/(KBLSTWSVM) is capable of generating two linear and non-parallel hyperplanes based on real data and prior knowledge. KBLSTWSVM retains the advantage of LSTWSVM formulations; solution can be obtained by solving two systems of linear equations. Further, exploiting the special structure of the matrix, partition method can be used to efficiently calculate the matrix inverse. Experimental results on several benchmark datasets demonstrate the effectiveness of the proposed approach.

Before we present the formulations of KBTWSVM and KBLTWSVM, we introduce the formulation of KBPSVM.

7.3.1 Knowledge Based Proximal SVM

Khemchandani et al. [18] proposed KBPSVM, where they incorporated prior knowledge, represented by polyhedral sets into linear PSVM. Consider the problem of binary classification wherein a linearly inseparable data set of m points in real n-dimensional space of features is represented by the matrix $X \in \mathbb{R}^{m\times n}$. The corresponding target or class of each data point X_i; $i = 1, 2, \dots, m$, is represented by a diagonal matrix $D \in \mathbb{R}^{m\times n}$ with entries D_{ii} as $+1$ or -1. Let us assume that prior information in the form of the following knowledge sets is given for the two classes.

$$l_i \text{ sets belonging to class } +1 : \{x \mid H^i x \leq h^i\}, i = 1, 2, \dots, l_t,$$

$$k_i \text{ sets belonging to class } -1 : \{x \mid G^i x \leq g^i\}, i = 1, 2, \dots, K_t\,,$$

where H^i and G^i are $(1 \times n)$ row vectors; and h^i and $g^i \in \mathbb{R}^n$. Given the above problem, the aim of KBPSVM is to determine the hyperplane described by $w^T x + b = 0$, which lies midway between the two parallel proximal hyperplanes given by $w^T x + b = +1$ and $w^T x + b = -1$. KPSVM obtains this hyperplane by solving the following optimization problem.

$$\underset{(w,\,b)}{\text{Min}} \; \frac{1}{2}(w^T w + b^2) + \frac{c_1}{2} \parallel q \parallel^2 + \frac{c_2}{2}(\parallel r_1 \parallel^2 + r_2^2) + \frac{c_3}{2}(\parallel s_1 \parallel^2 + s_2^2) + \frac{c_4}{2}(\parallel u \parallel^2 + \parallel v \parallel^2)$$

subject to

$$\begin{aligned} D(Xw + eb) + q &= e, \\ H^t u + w &= r_1, \\ h^t u + b + 1 &= r_2, \end{aligned}$$

$$G^t v - w = s_1,$$
$$g^t v - b + 1 = s_2, \tag{7.17}$$

where $c_1, c_2, c_3, c_4 > 0$ are penalty parameters, q, r_1, r_2, s_1, s_2 are slack variables, and u, v are variables introduced in the process of incorporating polyhedral knowledge sets using Farkas theorem (Mangasarian [17]). It is to be noted that the proposed KBPSVM formulation considers only one knowledge set corresponding to each class $\{x \mid Hx \leq h\}$ and $\{x \mid Gx \leq g\}$ such that matrices $H \in \mathbb{R}^{l\times n}$ and $G \in \mathbb{R}^{k\times n}$; generalization to multiple knowledge sets is straight forward. It is worth mentioning that KBPSVM formulation uses 2-norm approximation of knowledge sets and slack variables whereas the original KBSVM formulation considered 1-norm approximation, both for the reason of computational advantage. Khemchandani et al. [18] showed the solution of the QPP (7.17) can be obtained by solving the following system

$$\widehat{U}\begin{bmatrix}\alpha\\ \beta\\ \sigma\end{bmatrix} = \begin{bmatrix}e\\ -e_1\\ -e_1\end{bmatrix}, \tag{7.18}$$

where

$$\widehat{U} = \begin{bmatrix} DPP^TD + \dfrac{I}{c_1} & DP & -DP \\ P^TD & \dfrac{I}{c_4}MM^T + (1+\dfrac{I}{c_2})I & -I \\ -P^TD & -I & \dfrac{I}{c_4}NN^T + (1+\dfrac{I}{c_3})I \end{bmatrix},$$

and $z = \begin{bmatrix} w\\ b\end{bmatrix} = P^TD\alpha + \beta - \sigma, \quad P = \begin{bmatrix}X & e\end{bmatrix}, \quad M = \begin{bmatrix}H & h\end{bmatrix}^T, \quad N = \begin{bmatrix}G & g\end{bmatrix}^T,$

$e_1 = \begin{bmatrix}0e\\ 1\end{bmatrix}_{(n+1)\times 1}, \ \alpha \in \mathbb{R}^m, \ \beta \in \mathbb{R}^{n+1}, \ \sigma \in \mathbb{R}^{n+1}.$

Thus, the solution of KBPSVM can be obtained by solving single system of linear equations of order $(m + 2(n + 1))$. However using matrix partition method and Sherman-Morrison-Woodbury (SMW) formula (Golub and Van Loan [19]) an efficient way of computing the desired solution has also been proposed in Khemchandani [18]. This solution requires three small inverses of order $(n + 1)$ each, instead of a single large $(m + 2(n + 1))$ matrix inverse. This solution is also advantageous in the sense that the order of the matrix to be inverted is independent of the number of data points m and depends only on the dimension n. This improves the training speed of KBPSVM significantly as usually $n << m$. Khemchandani et al. [18] have also empirically compared the proposed KBPSVM approach against KBSVM with 1-norm (KBSVM 1), KBSVM with 2-norm (KBSVM 2), conventional SVM and PSVM on two selected data sets WPBC and Cleveland Heart respectively.

Before proceeding to KBTSVM/KBLSTSVM, we will show an alternate way of solving *QPP* (7.17), which may turn-out to be advantageous in many cases. The solution of the system (7.18) was obtained in Khemchandani et al. [18] by constructing the Lagrangian function and K.K.T conditions; here we will derive an alternate solution by direct substitution. To obtain the alternate solution we will substitute all the constraints of *QPP* (7.17) into the objective function. This will lead us to an unconstrained QPP as shown below

$$\begin{aligned}\text{Min}\quad & \frac{1}{2}(w^Tw+b^2)+\frac{c_1}{2}\parallel D(Xw+eb)-e\parallel^2+\frac{c_2}{2}(\parallel H^Tu+w\parallel^2+(h^Tu+b+1)^2)\\ & +\frac{c_3}{2}(\parallel G^Tv-w\parallel^2+(g^Tv-b+1)^2)+\frac{c_4}{2}(\parallel u\parallel^2+\parallel v\parallel^2).\end{aligned} \tag{7.19}$$

Setting the gradient of (7.19) with respect to w, b, u and v to zero gives the following equations

$$\begin{aligned}w+c_1A^TD[D(Xw+eb)-e]+c_2[H^Tu+w]-c_3[G^Tv-w]&=0e,\\ b-c_1e^TD[D(Xw+eb)-e]+c_2[h^Tu+b+1]-c_3[g^Tv-b+1]&=0,\\ c_2H[H^Tu+w]+c_2h[h^Tu+b+1]+c_4u&=0e,\\ c_3G[G^Tv-w]+c_3[g^Tv-b+1]+c_4v&=0e.\end{aligned}$$

Let us rearrange them in matrix format to yield the alternate solution. For this let columns $Col1,\ Col2,\ Col3$ and $Col4$ be denoted as follows

$$Col1=\begin{pmatrix}(1+c_2+c_3)I+c_1X^TX\\ -c_1e^TX\\ c_2H\\ -c_3G\end{pmatrix},\qquad Col2=\begin{pmatrix}-c_1X^Te\\ (1+c_2+c_3+c_1e^Te)\\ c_2h\\ -c_3g\end{pmatrix},$$

$$Col3=\begin{pmatrix}c_2H^T\\ c_2h^T\\ c_4+c_2hh^T+c_2HH^T\\ 0ee^T\end{pmatrix},\qquad Col4=\begin{pmatrix}-c_3G^T\\ -c_3g^T\\ 0ee^T\\ c_4+c_3gg^T+c_3GG^T\end{pmatrix}.$$

Then

$$\begin{bmatrix}Col1,\ Col2,\ Col3,\ Col4\end{bmatrix}\begin{bmatrix}w\\ b\\ u\\ v\end{bmatrix}=\begin{bmatrix}c_1X^TDe\\ -(c_1e^TDe+c_2-c_3)\\ -c_2h\\ -c_3g\end{bmatrix}. \tag{7.20}$$

This solution requires solving a single system of linear equations of order $n+1+l+k$. As before, using matrix partition method will lead us to an effective solution which requires three smaller inverses of order $(n+1)$, l and k each. It is to be noted that the solution (7.18) proposed in Khemchandani et al. [18] also requires three inverses of order $(n + 1)$, the advantage in using (7.20) is that often in practical cases $l, k << n$. The advantage is apparent particularly under the scenario of increasing number of knowledge sets. For example, consider including a new class +1 knowledge set $\{x| H_1 x \leq h_1\}, H_1 \in \mathbb{R}^{l_1 \times n}$ into KBPSVM formulation (7.17). Then, the new solution of type (7.18) will require and additional inverse of order $(n + 1)$ together with the original three inverses of order $(n + 1)$. Solution of type (7.20) will need an additional inverse of order l_1 together with the original three inverses. In other words, for every additional knowledge set included into the KBPSVM formulation, solution (7.18) requires same number of additional inverses of order $(n + 1)$, however solution (7.20) will need additional inverses whose order will depend upon the number of constraints in the additional knowledge set (which is usually lesser than n). We will describe our KBLSTSVM algorithm in the light of this alternate solution.

Now, let us consider the problem of incorporating two knowledge sets: $\{x \mid Hx \leq h\}$ belonging to class +1 and $\{x \mid Gx \leq g\}$ belonging to class −1 ($H \in \mathbb{R}^{l \times n}, h \in \mathbb{R}^l, G \in \mathbb{R}^{k \times n}$ and $g \in \mathbb{R}^k$) into TWSVM formulation; generalization to multiple knowledge sets is straight forward. First, we will discuss incorporation of these knowledge sets into $(TWSVM1)$; incorporation into $(TWSVM2)$ will be on similar lines. The idea here is to use the fact that in each QPP of TWSVM, the objective function corresponds to a particular class and the constraints are determined by patterns of the other class. Following Fung et al. [16] and Khemchandani et al. [18], the knowledge set $\{x \mid Gx \leq g\}$ can be incorporated into QPP for $(TWSVM1)$ as

$$
\begin{aligned}
&\underset{(w_1,\, b_1)}{\text{Min}} \quad \frac{1}{2} \| Aw_1 + eb_1 \|^2 + c_1 e^T q_1 + d_1 (e^T r_1 + p_1) \\
&\text{subject to} \\
&\qquad -(Bw_1 + eb_1) + q_1 \geq e, \quad q_1 \geq 0e, \\
&\qquad -r_1 \leq G^T u_1 - w_1 \leq r_1, \quad u_1 \geq 0e, \quad r_1 \geq 0e, \\
&\qquad g^T u_1 - b_1 + 1 \leq p_1, \quad p_1 \geq 0.
\end{aligned} \tag{7.21}
$$

Incorporating knowledge set $\{x \mid Gx \leq g\}$ corresponding to class −1 into QPP $(TWSVM2)$ directly follows (Fung et al. [16] and Khemchandani et al. [18]), as $(TWSVM1)$ treats class −1 data points in the same way as conventional SVM. However, for incorporating knowledge set $\{x \mid Hx \leq h\}$ of class +1, we should understand the fact for $(TWSVM1)$, the QPP requires the hyperplane $x^T w_1 + b_1 = 0$ to be closer to class +1 data points. Hence, the hyperplane $x^T w_1 + b_1 = 0$ should be closer to the knowledge sets of class +1 as well. Hence we will impose the condition that, for all data points $\{x, Hx \leq h\}$, $\| x^T w_1 + b_1 \|_1$ should be minimum (here we are using 1-norm instead of standard 2-norm for the reason of simplicity). This is equivalent to stating that for all datapoints $\{x \mid Hx \leq h\}$, z_1 should be minimum, where $-z_1 \leq (x^T w_1 + b_1) \leq z_1$. It is equivalent to the following statements (with the condition z_1 should be minimum) $Hx \leq h$ and $-z_1 > x^T w_1 + b_1$ has no solution x and $Hx \leq h$ and $x^T w_1 + b_1 > z_1$ has no solution x.

These above statements are implied by the following statements (Fung et al. [16]):

- $H^T v_1 + w_1 = 0e, h^T v_1 + b_1 - z_1 \leq 0, \ v_1 \geq 0e$, has a solution (v_1, w_1) and
- $H^T t_1 - w_1 = 0e, h^T t_1 - b_1 - z_1 \leq 0, \ t_1 \geq 0e$, has a solution $(t_1, \ w_1)$.

Incorporating these conditions into QPP (7.21) and adding slack variables s_1 and y_1 will yield the final knowledge based version of QPP (7.21) as

$$\underset{(w_1,b_1)}{\text{Min}} \quad \frac{1}{2} \parallel Aw_1 + eb_1 \parallel^2 + c_1 e^T q_1 + d_1(e^T r_1 + p_1) + f_1(z_1 + e^T s_1 + e^T y_1)$$

subject to

$$\begin{aligned} -(Bw_1 + eb_1) + q_1 \geq e, \quad q_1 \geq 0e, \\ -r_1 \leq G^T u_1 - w_1 \leq r_1, \quad u_1 \geq 0e, \quad r_1 \geq 0e, \\ g^T u_1 - b_1 + 1 \leq p_1, \quad p_1 \geq 0e, \\ -s_1 \leq H^T v_1 + w_1 \leq s_1, \quad v_1 \geq 0e, \quad s_1 \geq 0e, \\ h^T v_1 + b_1 \leq z_1, \\ -y_1 \leq H^T t_1 - w_1 \leq y_1, \quad t_1 \geq 0e, \quad y_1 \geq 0e, \\ h^T t_1 - b_1 \leq z_1, \quad z_1 \geq 0. \end{aligned} \tag{7.22}$$

In an exactly similar way knowledge sets can be incorporated into QPP for (*TWSVM2*) and the following modified QPP can be obtained

$$\underset{(w_2,\ b_2)}{\text{Min}} \quad \frac{1}{2} \parallel Bw_2 + eb_2 \parallel^2 + c_2 e^T q_2 + d_2(e^T r_2 + p_2) + f_2(z_2 + e^T s_2 + e^T y_2)$$

subject to

$$\begin{aligned} Aw_2 + eb_2 + q_2 \geq e, \quad q_2 \geq 0e, \\ -r_2 \leq H^T u_2 - w_2 \leq r_2, \quad u_2 \geq 0e, \quad r_2 \geq 0e, \\ h^T u_2 + b_2 + 1 \leq p_2, \quad p_2 \geq 0, \\ -s_2 \leq G^T v_2 + w_2 \leq s_2, \quad v_2 \geq 0e, \quad s_2 \geq 0e, \\ g^T v_2 + b_2 \leq z_2, \\ -y_2 \leq G^T t_2 - w_2 \leq y_2, \quad t_2 \geq 0e, \quad y_2 \geq 0e, \\ g^T t_2 - b_2 \leq z_2, \quad z_2 \geq 0. \end{aligned} \tag{7.23}$$

Together, QPPs (7.22) and (7.23) define our final KBTSVM formulation, which is capable of producing linear non-parallel hyperplanes $w_1^T x + b_1 = 0$ and $w_2^T x + b_2 = 0$ from real data (A, B) and prior knowledge $\{x \mid Hx \leq h\}, \ \{x \mid Gx \leq g\}$. Here $c_1, c_2, d_1, d_2, f_1, f_2 > 0$ are trade-off parameters and $q_1, q_2, r_1, r_2, p_1, p_2, z_1, z_2, s_1, s_2, y_1, y_2$ are slack variables and $u_1, u_2, v_1, v_2, t_1, t_2$ are variables introduced in the process of incorporating polyhedral knowledge sets using Farkas Theorem (Mangasarian [17]). It is easy to see that our KBTWSVM formulation simplifies to TWSVM under the condition $d_1, d_2, f_1, f_2 = 0$. Also, if we would like to give equal significance to data and prior knowledge we can set $d_1 = c_1, d_2 = c_2$, and $f_1, f_2 = 1$.

Following Jayadeva et al. [20], dual QPPs of (7.22) and (7.23) can be derived to obtain the solution of KBTWSVM. However, considering the fact that LSTWSVM has been shown to be significantly faster in training than TWSVM without any compromise in accuracy (Kumar et al. [3]), we will use the above KBTWSVM formulation to obtain KBLSTSVM in the next section and proceed to computational experiments with KBLSTSVM in subsequent sections.

7.3.2 Knowledge Based Least Squares Twin SVM

In this section, following Fung et al. [16] and Kumar et al. [3] we will take equalities in primal QPPs (7.22) and (7.23) instead of inequalities to get our KBLSTSVM formulation. The new QPPs of KBLSTSVM formulation is shown in (7.24) and (7.25).

$$
\begin{aligned}
\underset{(w_1,b_1)}{\text{Min}} \quad & \frac{1}{2} \| Aw_1 + eb_1 \|^2 + \frac{c_1}{2} \| q_1 \|^2 + \frac{d_1}{2}(\| r_1 \|^2 + p_1^2 + \| u_1 \|^2) + \\
& \frac{f_1}{2}(z_1^2 + \| s_1 \|^2 + \| y_1 \|^2 + \| v_1 \|^2 + \| t_1 \|^2)
\end{aligned}
$$

subject to

$$
\begin{aligned}
-(Bw_1 + eb_1) + q_1 &= e, \\
G^T u_1 - w_1 &= r_1, \\
g^T u_1 - b_1 + 1 &= p_1, \\
H^T v_1 + w_1 &= s_1, \\
h^T v_1 + b_1 &= z_1, \\
H^T t_1 - w_1 &= y_1, \\
h^T t_1 - b_1 &= z_1.
\end{aligned} \tag{7.24}
$$

and

$$
\begin{aligned}
\underset{(w_2,\, b_2)}{\text{Min}} \quad & \frac{1}{2} \| Bw_2 + eb_2 \|^2 + \frac{c_2}{2} \| q_2 \|^2 + \frac{d_2}{2}(\| r_2 \|^2 + p_2^2 + \| u_2 \|^2) + \\
& \frac{f_2}{2}(z_2^2 + \| s_2 \|^2 + \| y_2 \|^2 + \| v_2 \|^2 + \| t_2 \|^2)
\end{aligned}
$$

subject to

$$\begin{aligned} Aw_2 + eb_2 + q_2 &= e, \\ H^T u_2 + w_2 &= r_2, \\ h^T u_2 + b_2 + 1 &= p_2, \\ G^T v_2 + w_2 &= s_2, \\ g^T v_2 + b_2 &= z_2, \\ G^T t_2 - w_2 &= y_2, \\ g^T t_2 - b_2 &= z_2. \end{aligned} \tag{7.25}$$

By substituting the constraints into the objective function of QPP (7.24) we get

$$\begin{aligned} \underset{(w_1,\, b_1)}{\text{Min}} \; & \frac{f_1}{2}\left(\frac{1}{4}(h^T t_1 + h^T v_1)^2 + \| H^T v_1 + w_1 \|^2 + \| H^T t_1 - w_1 \|^2 + \| v_1 \|^2 + \| t_1 \|^2\right) \\ & + \frac{d_1}{2}(\| G^T u_1 - w_1 \|^2 + (g^T u_1 - b_1 + 1)^2 + \| u_1 \|^2) + \frac{1}{2} \| Aw_1 + eb_1 \|^2 \\ & + \frac{c_1}{2} \| -(Bw_1 + eb_1) + e \|^2 . \end{aligned} \tag{7.26}$$

Now on setting the gradient of above with respect to w_1, b_1, u_1, v_1 and t_1 to zero, we get the following equations:

$$\begin{aligned} A^T[Aw_1 + eb_1] - c_1 B^T[-(Bw_1 + eb_1) - e] - d_1[G^T u_1 - w_1] + & \\ f_1[H^T v_1 + w_1] - f_1[H^T t_1 - w_1] &= 0e, \\ -e^T[Aw_1 + eb_1] + c_1 e^T[-(Bw_1 + eb_1) - e] - d_1[g^T u_1 - b_1 + 1] &= 0, \\ d_1 G[G^T u_1 - w_1] + d_1 g[g^T u_1 - b_1 + 1] + d_1 u_1 &= 0e, \\ \frac{f_1}{4} h(h^T t_1 + h^T v_1) + f_1 H[h^t v_1 + w_1] + f_1 v_1 &= 0e, \\ \frac{f_1}{4} h(h^T t_1 + h^T v_1) + f_1 H[h^t t_1 - w_1] + f_1 t_1 &= 0e. \end{aligned}$$

Let

$$E = \begin{bmatrix} E_1 \; E_2 \; E_3 \; E_4 \; E_5 \end{bmatrix},$$

where

$$E_1 = \begin{bmatrix} A^T A + c_1 B^T B + I(d_1 + 2f_1) \\ -e^T A - c_1 e^T B \\ -d_1 G \\ f_1 H \\ -f_1 H \end{bmatrix} \qquad E_2 = \begin{bmatrix} [A^T e + c_1 B^T e] \\ m_1 - c_1 m_2 + d_1 \\ -d_1 g \\ 0e \\ 0e \end{bmatrix}$$

$$E_3 = \begin{bmatrix} -d_1 G^T \\ -d_1 g^T \\ d_1[gg^T + GG^T + I] \\ 0ee^T \\ 0ee^T \end{bmatrix} \qquad E_4 = \begin{bmatrix} f_1 H^T \\ 0e^T \\ 0ee^T \\ f_1[\frac{hh^T}{4} + HH^T + I] \\ \frac{f_1 hh^T}{4} \end{bmatrix},$$

$$E_5 = \begin{bmatrix} -f_1 H^T \\ 0e^T \\ 0ee^T \\ \frac{f_1 hh^T}{4} \\ f_1[\frac{hh^T}{4} + HH^T + I]. \end{bmatrix}$$

Then the solution of QPP (7.24) is

$$E \begin{bmatrix} w_1 \\ b_1 \\ u_1 \\ v_1 \\ t_1 \end{bmatrix} = \begin{bmatrix} -c_1 B^T e \\ d_1 + c_1 m_2 \\ -d_1 g \\ 0e \\ 0e \end{bmatrix}. \tag{7.27}$$

The square matrix E is of order $(n+1+k+2l) \times (n+1+k+2l)$. Let

$$F = \begin{bmatrix} F_1 \ F_2 \ F_3 \ F_4 \ F_5 \end{bmatrix},$$

where

$$F_1 = \begin{bmatrix} B^T B + c_2 A^T A + I(d_2 + 2f_2) \\ -[e^T B + c_2 e^T A] \\ -d_2 H \\ f_1 G \\ -f_1 G \end{bmatrix}, \qquad F_2 = \begin{bmatrix} -[B^T e + c_2 A^T e] \\ m_2 + c_2 m_1 + d_2 \\ -d_2 h \\ 0e \\ 0e \end{bmatrix},$$

$$F_3 = \begin{bmatrix} d_2 H^T \\ -d_2 h^T \\ d_2[hh^T + HH^T + I] \\ 0ee^T \\ 0ee^T \end{bmatrix}, \qquad F_4 = \begin{bmatrix} f_2 G^T \\ oe^T \\ oee^T \\ f_2[\frac{gg^T}{4} + GG^T + I] \\ \frac{f_2 gg^T}{4} \end{bmatrix},$$

$$F_5 = \begin{bmatrix} -f_2 g^T \\ oe^T \\ oee^T \\ \frac{f_2 gg^T}{4} \\ f_2[\frac{gg^T}{4} + GG^T + I] \end{bmatrix}.$$

In an exactly similar way the solution of QPP (7.25) can be derived to be

$$F \begin{bmatrix} w_2 \\ b_2 \\ u_2 \\ v_2 \\ t_2 \end{bmatrix} = \begin{bmatrix} -c_2 A^T e \\ -d_2 - c_2 m_1 \\ -d_2 h \\ 0e \\ 0e \end{bmatrix}. \tag{7.28}$$

The square matrix F is of order $(n+1+l+2k) \times (n+1+l+2k)$. Thus the solution of KBLSTSVM (w_1, b_1, w_2, b_2) can be obtained by solving two systems of linear equations (7.27) and (7.28). It is easy to see that by using matrix partition method the solutions of both (7.27) and (7.28) can be obtained by finding three smaller inverses of order $(n+1)$, l and k each. This solution is extremely simple when compared with solving two QPPs in KBSTSVM; thus KBLSTSVM retains the advantage of LSTWSVM over TWSVM. Once the solution is obtained, a new datapoint $x \in \mathbb{R}^n$ is assigned to a class $+1$ or -1 depending on to which of the two hyper planes, its perpendicular distance is minimum.

7.3.3 Computational Experiments

To demonstrate the performance of KBLSTSVM, several experiments on benchmark datasets were conducted in Fung et al. [16] and Khemchandani et al. [18]. We used conducted experiments on KBSVM1, KBSVM2, and KBPSVM for comparison with KBLSTSVM on the same datasets. All algorithms were implemented in MATLAB 7.3.0 (R2006b) environment on a PC with Intel Core2Duo processor (2.13 GHz), 1 GB RAM. The LPPs/QPPs of KBSVM1/KBSVM2 were solved using mosek optimization toolbox for MATLAB, which implements fast interior point based algorithms. Classification accuracy of each algorithm was measured by standard tenfold cross-validation methodology. For all algorithms we tuned the penalty parameters using a tuning set as done in Khemchandani et al. [18].

We use two datasets in our experiments. The first dataset used was a subset of Wisconsin breast cancer prognosis (WPBC) dataset consisting of 108 datapoints using a 60-month cutoff for predicting recurrence or non-recurrence of the disease. Following Khemchandani et al. [18] we used the prognosis rules used by doctors Lee et al. [21], as prior knowledge. These rules depend upon two features from the dataset

Table 7.6 Testing accuracy with linear kernel

Dataset	KBLSTWSVM	KBPSVM	KBSVM2	KBSVM1
WPBC (108 × 33)	69.43 ± 10.84	68.87 ± 9.7	67.87 ± 12.15	66.62 ± 15.54
Cleveland heart (303 × 13)	69.43 ± 10.84	69.43 ± 10.84	69.43 ± 10.84	69.43 ± 10.84

Table 7.7 Training times

Method	WPBC	Cleveland heart
KBSVM2	1.2196	0.5
KBSVM1	0.9819	0.311
KBLSTWSVM	1.09E-2	1.02E-2
KBPSVM	5.7E-3	6.1E-3

namely tumor size (T) (feature 31) and lymph node status (L) (feature 32). Tumor size is the diameter of the excised tumor in centimeters, and lymph node status refers to the number of metastasized auxiliary lymph nodes. The rules are given by:

$$(L \geq 5)(T \geq 4) \Rightarrow \text{Recur}, \ (L = 0)(T \leq 1.9) \Rightarrow \text{Non Recur}.$$

The second dataset used was Cleveland heart dataset (Blake and Merz [14]) using a less than 50 % cutoff diameter narrowing for predicting presence or absence of heart disease. The knowledge sets consists of, rules constructed (for the sake of illustration) and used in Fung et al. [16] and Khemchandani et al. [18] with two features oldpeak (feature 10) and thal (T) (feature 13). The rules are given by:

$$(T \geq 4)(O \geq 4) \Rightarrow \text{Presence of heart disease},$$

$$(T \leq 3)(O \leq 1) \Rightarrow \text{Absence of heart disease},$$

Table 7.6 shows the comparison of classification accuracy for KBLSTSVM, KBPSVM, KBSVM1 and KBSVM2 on chosen datasets. It could be observed that KBLSTSVM gives comparable performance on both WPBC and Cleveland Heart datasets. Table 7.7 compares the total training time taken by all these classifiers. While KBLSTSVM is fast when compared to KBSVMs, it is not as fast as KBPSVM. This is obvious as KBLSTSVM requires solving two sets of linear equations as opposed to one in KBPSVM.

7.4 Support Tensor Machines: A Brief Review

In the machine learning community, high dimensional image data with many attributes are often encountered in the real applications. The representation and selection of the features will have a strong effect on the classification performance. Thus, how to efficiently represent the image data is one of the fundamental problems in classifier model design. It is worth noting that most of existing classification algorithms are oriented to vector space model (VSM), e.g. support vector machine (SVM) (Cristianini and Shawe-Taylor [22]; Vapnik [23]), proximal support vector machine classifier (PSVM) (Fung and Mangasarian [13]), twin support vector machine for pattern classification (TWSVM) (Jayadeva et al. [20]).

SVM relies on a vector dataset, takes vector data x in space $\mathbb{R}^n$ as inputs, and aims at finding a single linear (or nonlinear) function. Thus, if SVM are applied for image classification, images are commonly represented as long vectors in the high-dimensional vector space, in which each pixel of the images corresponds to a feature or dimension. For example, when the VSM focused methods are applied and one is often confronted with an image with 64×64 pixels, then image x will be represented as a long vector $\hat{x}$ with dimension $n = 4{,}096$. In such cases, learning such a linear SVM function $w^T\hat{x} + b$ in vector space is time-consuming, where $w \in \mathbb{R}^n$ and b are parameters to be estimated. And most importantly, such a vector representation fails to take into account the spatial locality of pixels in the image (Wang et al. [24]; Zhang and Ye [25]).

Images are intrinsically matrices. To represent the images appropriately, it is important to consider transforming the vector patterns to the corresponding matrix patterns or second order tensors before classification. In recent years, some interests about tensor representation have been investigated. Specifically, some tensor representation based approaches (Fu and Huang [26]; He et al. [27]; Yan et al. [28]) are proposed for high-dimensional data analysis. A tensor-based learning framework for SVMs, termed as support tensor machines (STMs), was recently proposed by Cai et al. [4], and Tao et al. [29, 30], which directly accepts tensors as inputs in the learning model, without vectorization. Tao et al. [29, 30] have discussed tensor version of various classification algorithms available in the literature. The use of tensor representation in this way helps overcome the overfitting problem encountered mostly in vector-based learning.

7.4.1 Basics of Tensor Algebra

A Tensor T with order k is a real valued multilinear function on k vector spaces

$$T : \mathbb{R}^{n_1} \times \cdots \times \mathbb{R}^{n_k} \rightarrow \mathbb{R}. \tag{7.29}$$

The set of all k-tensors on $\{\mathbb{R}^{n_i}; i = 1, \ldots, k\}$, denoted by $\mathcal{T}^k$, is a vector space under the usual operations of pointwise addition and scalar multiplication:

$$(aT)(a_1, \ldots, a_k) = a(T(a_1, \ldots, a_k)), \tag{7.30}$$

$$(T + T^{'})(a_1, \ldots, a_k) = T(a_1, \ldots, a_k) + T^{'}(a_1, \ldots, a_k), \tag{7.31}$$

where $a_i \in \mathbb{R}^{n_i}$. Given two tensors $S \in T^k$ and $T \in T^l$, define a map

$$S \otimes T : \mathbb{R}^{n_1} \times \cdots \times \mathbb{R}^{n_{k+l}} \to \mathbb{R}, \tag{7.32}$$

by

$$S \otimes T(a_1, \ldots, a_{k+l}) = S(a_1, \ldots, a_k)T(a_{k+1}, \ldots, a_{k+l}), \tag{7.33}$$

such that $S \otimes T$ is a $(k+l)$-tensor, called the tensor product of S and T.

Let $e_1, \ldots, e_{n_1}$ be the standard basis of $\mathbb{R}^{n_1}$, and $\epsilon_1, \ldots, \epsilon_{n_1}$ be the dual basis of $\mathbb{R}^{n_1}$ which is formed by coordinate functions with respect to the basis of $\mathbb{R}^{n_1}$. Similarly, let $\hat{e_1}, \ldots, \hat{e_{n_1}}$ be the standard basis of $\mathbb{R}^{n_2}$, and $\hat{\epsilon_1}, \ldots, \hat{\epsilon_{n_1}}$ be the dual basis of $\mathbb{R}^{n_2}$. We have

$$\epsilon_i(e_j) = \delta_{ij} \quad \text{and} \quad \hat{\epsilon}_i(\hat{e}_j) = \delta_{ij} \tag{7.34}$$

where δ_{ij} is the Kronecker Delta function. Thus, $\{\epsilon_i \otimes \hat{\epsilon}_j\}$ $(1 \le i \le n_1,\ 1 \le j \le n_2)$ forms a basis of $\mathbb{R}^{n_1} \otimes \mathbb{R}^{n_2}$. For any order-2 tensor T, we can write it as

$$T = \sum_{\substack{1 \le i \le n_1, \\ 1 \le j \le n_2}} T_{ij} \epsilon_i \otimes \hat{\epsilon}_j. \tag{7.35}$$

This shows that every order-2 tensor in $R^{n_1} \otimes R^{n_2}$ uniquely corresponds to an $n_1 \times n_2$ matrix. Given two vectors $a = \sum_{k=1}^{n_1} a_k e_k \in \mathbb{R}^{n_1}$ and $b = \sum_{l=1}^{n_2} b_l \hat{e}_l \in \mathbb{R}^{n_2}$, we have

$$\begin{aligned} T(a, b) &= \sum_{i,j} T_{ij} \epsilon_i \otimes \hat{\epsilon}_j \left(\sum_{k=1}^{n_1} a_k e_k, \sum_{l=1}^{n_2} b_l \hat{e}_l \right) \\ &= \sum_{i,j} T_{ij} \epsilon_i \left(\sum_{k=1}^{n_1} a_k e_k \right) \hat{\epsilon}_j \left(\sum_{l=1}^{n_2} b_l \hat{e}_l \right) \\ &= \sum_{i,j} T_{ij} a_i b_j \\ &= a^T T b. \end{aligned} \tag{7.36}$$

7.4.2 *Tensor Space Model*

Different from Vector Space Model which considers input in the form of a vector in $\mathbb{R}^n$, Tensor Space Model considers the input variable as a second order tensor in $\mathbb{R}^{n_1} \otimes \mathbb{R}^{n_2}$, where $n_1 \times n_2 \approx n$. A linear classifier in $\mathbb{R}^n$ is represented as $a^T x + b$, in which there are $n+1$ $(\approx n_1 \times n_2 + 1)$ parameters $(b, a_i, i = 1 \ldots n)$. Similarly, a linear classifier in the tensor space $\mathbb{R}^{n_1} \otimes \mathbb{R}^{n_2}$ can be represented as $u^T X v + b$ where $u \in \mathbb{R}^{n_1}$ and $v \in \mathbb{R}^{n_2}$. Thus, there are only $n_1 + n_2 + 1$ parameters to be estimated.

This property makes tensor-based learning algorithms especially suitable for small sample-size (S3) cases.

Problem Statement

Given a set of training samples $\{(X_i, y_i); i = 1, \ldots, m\}$, where X_i is the data point in order-2 tensor space, $X_i \in \mathbb{R}^{n_1} \otimes \mathbb{R}^{n_2}$ and $y_i \in \{-1, 1\}$ is the class associated with X_i, STM finds a tensor classifier $f(X) = u^T X v + b$ such that the two classes can be separated with maximum margin.

Algorithm

STM is a tensor generalization of SVM. The algorithmic procedure is formally stated below

Step 1 Initialization: Let $u = (1, \ldots, 1)^T$.

Step 2 Computing v: Let $x_i = {X_i}^T u$ and $\beta_1 = ||u||^2$, v can be computed by solving the following optimization problem

$$\begin{aligned} \underset{(v,b,\xi)}{\text{Min}} \quad & \frac{1}{2}\beta_1 v^T v + Ce^T \xi \\ \text{subject to} \quad & \\ & y_i(v^T x_i + b) \geq 1 - \xi_i, \\ & \xi_i \geq 0, \qquad (i = 1, \ldots, m). \end{aligned}$$

where e is a vector of ones of appropriate dimension, and $\xi = (\xi_1, \ldots, \xi_m)$ is the slack vector.

Step 3 Computing u: Let $\tilde{x}_i = X_i v$ and $\beta_2 = ||v||^2$. Once v is obtained, u can then be computed by solving the following optimization problem

$$\begin{aligned} \underset{(u,b,\xi)}{\text{Min}} \quad & \tfrac{1}{2}\beta_2 u^T u + Ce^T \xi \\ \text{subject to} \quad & \\ & y_i(u^T \tilde{x}_i + b) \geq 1 - \xi_i, \\ & \xi_i \geq 0, \qquad (i = 1, \ldots, m). \end{aligned}$$

Step 4 Iteratively computing u and v: By steps 2 and 3, we can iteratively compute u and v until they tend to converge i.e. $\| u_i - u_{i-1} \| < \epsilon$ or $\| v_i - v_{i-1} \| < \epsilon$.

Theoretical Derivation

Suppose we have a set of order-2 tensors $X_1, \ldots, X_m \in \mathbb{R}^{n_1} \otimes \mathbb{R}^{n_2}$. A linear classifier in the tensor space can then be naturally represented as follows

$$f(X) = sign(u^T X v + b), \quad u \in \mathbb{R}^{n_1}, \quad v \in \mathbb{R}^{n_2}, \tag{7.37}$$

which can be rewritten through matrix inner product as follows

$$f(X) = sign(< X, uv^T > +b), \quad u \in \mathbb{R}^{n_1}, \quad v \in \mathbb{R}^{n_2}. \tag{7.38}$$

The optimization problem in the tensor space is then reduced to the following

$$\text{(TSO)} \qquad \underset{(u,v,b,\xi)}{\text{Min}} \quad \frac{1}{2}\|uv^T\|^2 + Ce^T\xi$$
subject to
$$y_i(u^T X_i v + b) \geq 1 - \xi_i,$$
$$\xi_i \geq 0, \quad (i = 1 \ldots m). \tag{7.39}$$

where e is a vector of ones of appropriate dimension. Let α and μ be the required Lagrangian variables. The Lagrangian for (7.39) is then given by:

$$L = \frac{1}{2}\|uv^T\|^2 + Ce^T\xi - \sum_i \alpha_i y_i(u^T X_i v + b) + e^T\alpha - \alpha^T\xi - \mu^T\xi.$$

Since

$$\begin{aligned}\frac{1}{2}\|uv^T\|^2 &= \frac{1}{2}\text{trace}(uv^T vu^T)\\ &= \frac{1}{2}(v^T u)\text{trace}(uu^T)\\ &= \frac{1}{2}(v^T v)(u^T u),\end{aligned}$$

we have

$$L = \frac{1}{2}(v^T v)(u^T u) + Ce^T\xi - \sum_i \alpha_i y_i(u^T X_i v + b) + e^T\alpha - \alpha^T\xi - \mu^T\xi.$$

Applying the K.K.T sufficient optimality conditions (Mangasarian [17] and Chandra [31]), we obtain

$$u = \frac{\sum_i \alpha_i y_i X_i v}{v^T v}, \tag{7.40}$$

$$v = \frac{\sum_i \alpha_i y_i u^T X_i}{u^T u}, \tag{7.41}$$

$$\alpha^T y = 0, \tag{7.42}$$

$$C - \alpha_i - \mu_i = 0, \quad (i = 1, \ldots, m). \tag{7.43}$$

From Equations (7.40) and (7.41), we see that u and v are dependent on each other, and cannot be solved independently. In the sequel, we describe a simple yet effective computational method to solve this optimization problem.

We first fix u. Let $\beta_1 = \|u\|^2$ and $x_i = X_i^T u$. The optimization problem (TSO) can be rewritten as follows

$$\begin{aligned}&\underset{(v,\, b,\, \xi)}{\text{Min}} \quad \frac{1}{2}\beta_1 \|v\|^2 + Ce^T\xi \\ &\text{subject to} \\ &\qquad y_i(v^T x_i + b) \geq 1 - \xi_i, \\ &\qquad \xi_i \geq 0, \quad (i = 1 \dots m).\end{aligned} \tag{7.44}$$

The new optimization problem is identical to the standard SVM. Once v is obtained, let $\tilde{x_i} = X_i v$ and $\beta_2 = \|v\|^2$, u can then be computed by solving the following optimization problem

$$\begin{aligned}&\underset{(u,\, b,\, \xi)}{\text{Min}} \quad \frac{1}{2}\beta_2 \|u\|^2 + Ce^T\xi \\ &\text{subject to} \\ &\qquad y_i(u^T \tilde{x_i} + b) \geq 1 - \xi_i, \\ &\qquad \xi_i \geq 0, \quad (i = 1 \dots m).\end{aligned} \tag{7.45}$$

Thus u and v can be obtained by iteratively solving the optimization problems (7.44) and (7.45).

Proof of Convergence

Define

$$f(u, v) = \frac{1}{2}\|uv^T\|^2 + Ce^T\xi, \tag{7.46}$$

where ξ is the slack vector and e is a vector of ones of appropriate dimension. Let u_0 be the initial value. Fixing u_0, we get v_0 by solving the optimization problem (7.44). Likewise, fixing v_0, we get u_1 by solving the optimization problem (7.45).

Since the optimization problem of SVM is convex, so the solution of SVM is globally optimum. Specifically, the solutions of problems (7.44) and (7.45) are globally optimum. Thus, we have

$$f(u_0, v_0) \geq f(u_1, v_0).$$

Hence we obtain the following chain of inequalities

$$f(u_0, v_0) \geq f(u_1, v_0) \geq f(u_1, v_1) \geq f(u_2, v_1) \geq \cdots .$$

Since f is bounded from below by zero, f converges.

Now the natural question is: why should a linear classifier in the tensor space $\mathbb{R}^{n_1} \times \mathbb{R}^{n_2}$ be of the form $u^T X v + b$ with $u \in \mathbb{R}^{n_1}$ and $v \in \mathbb{R}^{n_2}$?

To answer the above question, let us informally write a possible formulation of linear STM which is analogous to the formulation of linear SVM. As the data set is in the terms of matrices X_i having class labels y_i, $(i = 1, \ldots, m)$, it makes sense to determine a matrix $W \in \mathbb{R}^{n_1} \times \mathbb{R}^{n_2}$ and $b \in \mathbb{R}$ so that the linear STM classifier $< W, X > +b = 0$ is obtained by solving the following optimization problem

$$\begin{aligned}
&\underset{(W,b,\xi)}{\text{Min}} \quad \frac{1}{2} \parallel W \parallel_F^2 + C\sum_{k=1}^{n} \xi_k \\
&\text{subject to} \\
&\qquad y_k(< W, X_k > +b) \geq 1 - \xi_k, \quad (k = 1, \ldots, n), \\
&\qquad \xi_k \geq 0, \quad (k = 1, \ldots, n).
\end{aligned} \tag{7.47}$$

Here $C > 0$ is a trade-off parameter and $\xi_k (k = 1, \ldots, n)$ are slack variables. In problem (7.47), $< W, X >$ denotes the inner product of matrices W and X and is given by $< W, X >= \text{Trace}(WX^T)$. Further $\parallel W \parallel_F$ is the Frobenius norm which is given by $\parallel W \parallel_F^2 = \sum_i^{n_1} \sum_j^{n_2} | w_{ij} |^2$, i.e., $\parallel W \parallel_F = \sqrt{\text{Trace}(WW^T)}$.

Now using singular value decomposition of $W \in \mathbb{R}^{n_1 \times n_2}$ we can write

$$W = U \begin{pmatrix} \sum & 0 \\ 0 & 0 \end{pmatrix} V^T,$$

where $U \in \mathbb{R}^{n_1 \times n_1}$ and $V \in \mathbb{R}^{n_2 \times n_2}$ are orthogonal matrices, $\sum = diag(\sigma_1^2, \sigma_2^2, \ldots, \sigma_r^2)$, $\sigma_1 \geq \sigma_2, \geq, \ldots, \sigma_r > 0$ and $r = \text{Rank}(W)$. Denoting the matrices U and V by columns as $U = [\overline{u}_1, \overline{u}_2, \ldots, \overline{u}_{n_1}]$ and $V = [\overline{v}_1, \overline{v}_2, \ldots, \overline{v}_{n_2}]$, $\overline{u}_i \in \mathbb{R}^{n_1}$, $(i = 1, 2, \ldots, n_1)$ and $\overline{v}_j \in \mathbb{R}^{n_2}$, $(j = 1, 2, \ldots, n_2)$ we get

$$W = \sum_{i=1}^{r} \sigma_i^2 \overline{u}_i \overline{v}_j^T.$$

If we now take $u_i = \sigma_i \overline{u}_i$ and $v_i = \sigma_i \overline{v}_i$, we get

$$W = \sum_{i=1}^{r} u_i v_i^T,$$

which from (7.47) gives

$$\begin{aligned}
&\underset{(u_i,\, v_i,\, b,\, \xi)}{\text{Min}} \quad \frac{1}{2} \sum_{i=1}^{r} (u_i^T u_i)(v_i^T v_i) + C \sum_{k=1}^{n} \xi_K \\
&\text{subject to}
\end{aligned}$$

$$y_k \sum_{i=1}^{r} (u_i^T X_k v_i + b) \geq 1 - \xi_k, \quad (k = 1, \ldots, n),$$
$$\xi_k \geq 0, \quad (k = 1, \ldots, n). \quad (7.48)$$

Taking $r = 1$, we get $W = uv^T$, $u \in \mathbb{R}^{n_1}$, $v \in \mathbb{R}^{n_2}$ and therefore the problem (7.48) reduces to

$$\underset{(u,\ v,\ b,\ \xi)}{\text{Min}} \quad \frac{1}{2} \parallel uv^T \parallel_F + C \sum_{i=1}^{n} \xi_i$$

subject to

$$y_i(u^T X_i v + b) \geq 1 - \xi_i, \quad (i = 1, \ldots, n),$$
$$\xi_i \geq 0, \quad (i = 1, \ldots, n). \quad (7.49)$$

Most of the development in this area have concentrated only on Rank one SVM's. But recently some researchers have initiated work on Multiple Rank Support Tensor Machines, which have also been termed as Multiple Rank Support Matrix Machines for Matrix Data Classification. In this context the work of Gao et al. [32] looks promising.

The above discussion suggests that the decision function of STM is $f(x) = u^T X v + b$, where $X \in \mathbb{R}^{n_1} \times \mathbb{R}^{n_2}$, $b \in R$. If we now denote the Kronecker product operation (denoted by $\otimes$), then we can verify that

$$f(X) = Vec(u^T X v^T) + b = (v^T \otimes u^T) Vec(X) + b = (v \otimes u)^T Vec(X) + b.$$

Here $Vec(X)$ denotes an operator that vectorizes a matrix $X \in \mathbb{R}^{n_1} \times \mathbb{R}^{n_2}$ into a vector having $(n_1 \times n_2)$ components $(x_{11}, \ldots, x_{1n_2}; \ldots; x_{n_1 1}, \ldots, x_{n_1 n_2})$. Now comparing with SVM classifier $f(x) = w^T x + b$, $x \in \mathbb{R}^n$ and STM classifier $f(x) = u^T x v + b$ as obtained at (7.48) it follows that the decision functions in both SVM and STM have the same form and $v \otimes u$ in STM plays the same role as the vector w in SVM.

7.4.3 Proximal Support Tensor Machines

In a spirit similar to PSVM (Fung and Mangasarian [13]), Khemchandani et al. [2] developed a least squares variant of STM in the tensor space model, which classifies a new point to either of the two classes depending on its closeness to one of the two parallel hyperplanes in the tensor space. By using a least squares error term in the objective function, we reduce the inequality constraints of STM to equality constraints, and proceeding along the lines of STM, we develop an iterative algorithm which requires the solution of a system of linear equations at every iteration, rather than solving a dual QPP at every step, as in the case of STM. This leads to the formation of a fast tensor-based iterative algorithm, which requires the determination of a significantly smaller number of decision variables as compared to vector-based approaches. The above properties make (PSTMs) specially suited for overcoming overfitting in Small Sample Size (S3) problems.

Given a set of training samples $\{(X_i, y_i); i = 1, \ldots, m\}$, where X_i is the data point in order-2 tensor space, $X_i \in \mathbb{R}^{n_1} \otimes \mathbb{R}^{n_2}$ and $y_i \in \{-1, 1\}$ is the class associated with X_i, find a tensor classifier $f(X) = u^T X v + b$ in the linear tensor space, such that the two classes are proximal to either of the two parallel hyperplanes in the tensor space, separated maximally using margin maximization. Here, $u \in \mathbb{R}^{n_1}$, $v \in \mathbb{R}^{n_2}$, and $b \in \mathbb{R}$ are the decision variables to be determined by PSTM.

A linear classifier in the tensor space can be naturally represented as follows

$$f(X) = sign(u^T X v + b), \quad u \in \mathbb{R}^{n_1}, \quad v \in \mathbb{R}^{n_2}$$

which can be rewritten through matrix inner product as (Golub and Loan [19])

$$f(X) = sign(< X, uv^T > +b), \quad u \in \mathbb{R}^{n_1}, \quad v \in \mathbb{R}^{n_2}$$

The optimization problem of PSTM in the tensor space is given by:

$$\underset{\substack{(u,\ v,\ b,\ q_i) \\ i=1,\ldots,m}}{\text{Min}} \quad \frac{1}{2}||uv^T||^2 + \frac{1}{2}b^2 + \frac{1}{2}C\sum_{i=1}^{m} q_i{}^2$$

subject to

$$y_i(u^T X_i v + b) + q_i = 1, \quad (i = 1, \ldots, m), \tag{7.50}$$

where C is the trade-off parameter between margin maximization and empirical error minimization. Applying the K.K.T necessary and sufficient optimality conditions (Mangasarian [17] and Chandra [31]) to the problem (7.50), we obtain

$$u = \frac{\sum_{i=1}^{m} \alpha_i y_i X_i v}{v^T v}, \tag{7.51}$$

$$v = \frac{\sum_{i=1}^{m} \alpha_i y_i u^T X_i}{u^T u}, \tag{7.52}$$

$$b = \sum_{i=1}^{m} \alpha_i y_i, \tag{7.53}$$

$$q_i = \alpha_i / C, \quad (i = 1, \ldots, m). \tag{7.54}$$

We observe from (7.51) and (7.52) that u and v are dependent on each other, and cannot be solved independently. Hence, we resort to the alternating projection method for solving this optimization problem, which has earlier been used by Cai et al. [4] and Tao et al. [29] in developing STM. The method can be described as follows.

We first fix u. Let $\beta_1 = ||u||^2$ and $x_i = X_i{}^T u$. Let D be a diagonal matrix where $D_{ii} = y_i$. The optimization problem (7.50) can then be reduced to the following QPP

$$(v - PSTM) \qquad \underset{(v,b,q)}{\text{Min}} \quad \frac{1}{2}\beta_1 v^T v + \frac{1}{2}b^2 + \frac{1}{2}Cq^T q$$
$$\text{subject to}$$
$$D(Xv + eb) + q = e,$$

where $X = (x_1{}^T, x_2{}^T, \ldots, x_m{}^T)^T$, and e is a vector of ones of appropriate dimension. It can be seen that $(v - PSTM)$ is similar in structure to PSVM. Applying the K.K.T necessary and sufficient optimality conditions (Mangasarian [17]) to the problem $(v - PSTM)$, we obtain the following

$$\beta_1 v = X^T D\alpha, \tag{7.55}$$
$$b = e^T D\alpha, \tag{7.56}$$
$$Cq = \alpha. \tag{7.57}$$

Substituting the values of v, b and q from (7.55), (7.56) and (7.57) into the equality constraints of $(v - PSVTM)$ yields the following system of linear equations for obtaining α

$$\left[\frac{DXX^T D}{\beta_1} + Dee^T D + \frac{I}{C}\right]\alpha = e,$$

where I is an identity matrix of appropriate dimensions. Let $H_v = D[X/\sqrt{\beta_1} \quad e]$, and $G_v = \left[H_v H_v{}^T + I/C\right]$. Then,

$$\alpha = G_v{}^{-1} e.$$

Once α is obtained, v and b can be computed using (7.55) and (7.56) respectively.

We observe that solving $(v - PSTM)$ requires the inversion of an $m \times m$ matrix G_v, which is non-singular due to the diagonal perturbation introduced by the I/C term. Optionally, when $m \gg n_2$, we can instead use the (Golub and Van Loan [19]) to invert a smaller $(n_2 + 1) \times (n_2 + 1)$ matrix for obtaining the value of α, saving some computational time. The following reduction is obtained on applying the SMW identity (Golub and Loan [19]).

$$(H_v H_v{}^T + I/C)^{-1} = CI - C^2 H_v (I + CH_v{}^T I H_v)^{-1} H_v{}^T.$$

Once v is obtained, let $\tilde{x}_i = X_i v$ and $\beta_2 = ||v||^2$. u can then be computed by solving the following optimization problem:

$$(u - PSTM) \qquad \underset{(u,b,p)}{\text{Min}} \quad \frac{1}{2}\beta_2 u^T u + \frac{1}{2}b^2 + \frac{1}{2}Cp^T p$$
$$\text{subject to}$$
$$D(\tilde{X}u + eb) + p = e,$$

where $\tilde{X} = (\tilde{x_1}^T, \tilde{x_2}^T, \ldots, \tilde{x_m}^T)^T$. Solving $(u - PSTM)$ in a similar fashion as $(v - PSTM)$, we obtain the following equation for solving β

$$\beta = {G_u}^{-1} e,$$

where, $G_u = \left[H_u {H_u}^T + I/C\right]$ and $H_u = D[\tilde{X}/\sqrt{\beta_2} \quad e]$. Once β is obtained, u and b can be computed using the following equations:

$$\begin{aligned} \beta_2 u &= \tilde{X}^T D\beta, \\ b &= e^T D\beta. \end{aligned}$$

Thus, u and v can be obtained by iteratively solving the optimization problems $(v - PSTM)$ and $(u - PSTM)$. Since both the optimization problems are strictly convex, our proposed algorithm converges with a proof of convergence as described in the sequel.
Let

$$f(u, v) = \frac{1}{2}||uv^T||^2 + Ce^T\xi,$$

where ξ is the slack vector and e is a vector of ones of appropriate dimension. Let u_0 be the initial value. Fixing u_0, we first get v_0 by solving the optimization problem $(v - PSTM)$. Likewise, fixing v_0, we get u_1 by solving the optimization problem $(u - PSTM)$.

Since the optimization problem of PSVM is convex, the solutions of PSVM are globally optimum. Specifically, the solutions of equations $(v - PSTM)$ and $(u - PSTM)$ are globally optimum. Thus, we have:

$$f(u_0, v_0) \geq f(u_1, v_0).$$

Hence we obtain the following chain of inequalities:

$$f(u_0, v_0) \geq f(u_1, v_0) \geq f(u_1, v_1) \geq f(u_2, v_1) \geq \ldots.$$

Since f is bounded from below by zero, it converges.

We can now summarize the steps involved in solving PSTM using the alternating projection method as follows.

Algorithm for PSTM

Step 1 (Initialization)
Let $u = (1, \ldots, 1)^T$.

Step 2 (Computing v)
Let $\beta_1 = ||u||^2$, $x_i = {X_i}^T u$ and $X = ({x_1}^T, {x_2}^T, \ldots, {x_m}^T)^T$. v and b can then be determined by solving the following optimization problem

$$
\begin{array}{lll}
\text{(v-PSTM)} & \underset{(v,b,q)}{\text{Min}} & \frac{1}{2}\beta_1 v^T v + \frac{1}{2}b^2 + \frac{1}{2}Cq^T q \\
& \text{subject to} & \\
& & D(Xv + eb) + q = e.
\end{array}
$$

Step 3 (Computing u)
Once v is obtained, let $\tilde{x}_i = X_i v$, $\tilde{X} = (\tilde{x_1}^T, \tilde{x_2}^T, \ldots, \tilde{x_m}^T)^T$ and $\beta_2 = ||v||^2$. u can then be computed by solving the following optimization problem:

$$
\begin{array}{lll}
\text{(u-PSTM)} & \underset{(u,b,p)}{\text{Min}} & \frac{1}{2}\beta_2 u^T u + \frac{1}{2}b^2 + \frac{1}{2}Cp^T p \\
& \text{subject to} & \\
& & D(\tilde{X}u + eb) + p = e.
\end{array}
$$

Step 4 (Iteratively computing u and v)
If the iteration number exceeds the maximum number of iterations or all of the below convergence conditions are satisfied,

$$
\begin{aligned}
||u_i - u_{i-1}|| &\leq \text{tolerance}, \quad \text{and} \\
||v_i - v_{i-1}|| &\leq \text{tolerance}, \quad \text{and} \\
||b_i - b_{i-1}|| &\leq \text{tolerance},
\end{aligned}
$$

we stop our iterative algorithm, and u_i, v_i and b_i are the required optimal decision variables of PSTM. Iterate over steps 2 and 3 otherwise.

7.4.4 Experimental Results

Khemchandani et al. [2] compared the results of PSTM with another tensor-based classification method - STM on face detection and handwriting recognition datasets. In our presentation here, we restrict ourselves to hand writing data set only. For the face detection dataset and other details we refer to Khemchandani et al. [2]. Since this dataset is mostly balanced in both the cases, we consider percentage accuracy at varying sizes of training set as a performance metric for comparison (Duda et al. [15]; Vapnik [23]). For each size of the training set, we randomly select our training images from the entire set and repeat the experiment 10 times, in a spirit similar to (Cai et al. [4]; Hu et al. [33]). The mean and the standard deviation of the evaluation metrics are then used for comparison. In all our simulations, the initial parameters chosen for both PSTM and STM are: C = 0.5 and tolerance for convergence = 10^{-3}. All the algorithms have been implemented in MATLAB7 (R2008a) on Ubuntu9.04 running on a PC with system configuration Intel Core2 Duo (1.8 GHz) with 1 GB of RAM.

To provide a uniform testing environment to both PSTM and STM, the number of iterations performed by both the methods are taken to be the same for every simulation. Hence, for each train and test run, we first train the STM algorithm and obtain the number of iterations needed by STM to converge. This value is then used as the maximum number of iterations allowed for PSTM to converge on the same train and test run. After constraining the number of training iterations of PSTM in this fashion, computational time comparisons between PSTM and STM have also been reported.

There are two distinct handwriting recognition domains- online and offline, which are differentiated by the nature of their input signals (Plamondon and Srihari [34]). Online handwriting recognition systems depend on the information acquired during the production of the handwriting using specific equipments that capture the trajectory of the writing tool, such as in electronic pads and smart-phones. In this case, the input signal can be realized as a single-dimensional vector of trajectory points of the writing tool arranged linearly in time. On the other hand, in offline handwriting recognition systems, static representation of a digitized document such as cheque, form, mail or document processing is used. In this case, any information in the temporal domain regarding the production of the handwriting is absent, constraining the input signal to be handled as a binary matrix representation of images, where the ones correspond to the region in the image over which the character has been drawn by hand. Since, we are interested in working with tensor data, where the size of the input feature is large, we consider the offline recognition scenario, often called as Optical Character Recognition (OCR).

The Optdigits Dataset (Alpaydin and Kaynak [35]), obtained from the UCI repository, contains 32×32 normalized binary images of handwritten numerals 0–9, extracted from a preprinted form by using preprocessing programs made available by NIST. From a total of 43 people, 30 contributed to the training set and a different 13 to the test set, offering a wide variety of handwriting styles among the writers. The Optdigits dataset consists of 3 subsets of binary images - training data obtained from the first 30 subjects, writer-dependent testing data obtained from the same 30 subjects, and writer-independent testing data, which has been generated by the remaining 13 subjects and which is vital for comparing testing performances. The number of training and testing examples for each numeral is provided in Table 7.8.

Similar to the approach used in comparing performance over binary classifier, we consider a multi-class classification routine with varying ratio of the training set to be used. Also, the performance comparisons of PSTM and STM are evaluated on both the writer-dependent and writer-independent testing datasets, to study the effects of changing the writer on testing accuracy.

A Micro-Averaged Percentage Accuracy measure for different ratios of the training data is generated by solving $\binom{10}{2}$ binary classification problems. Result comparisons between PSTM and STM of this evaluation metric is provided in Table 7.9.

We further employ a one-against-one approach for extending the binary classifiers for multi-class classification of Opt Digits dataset. The class label for a new input test image is decided by a voting system across the $\binom{10}{2}$ binary classifiers. In this fashion, we obtain a macro-level comparison of percentage accuracy for multi-class

Table 7.8 Number of training and testing examples in optdigits dataset

Class	Number of training examples	Number of testing examples (Writer dependent)	Number of testing examples (Writer independent)
0	189	100	178
1	198	94	182
2	195	93	177
3	199	105	183
4	186	87	181
5	187	81	182
6	195	95	181
7	201	90	179
8	180	109	174
9	204	89	180

Table 7.9 Micro-averaged percentage accuracy comparisons on optdigits dataset

	Training ratio		
	0.01	0.3	0.7
Writer-independent			
PSTM	74.85 ± 26.37	86.81 ± 29.22	87.68 ± 29.47
STM	75.05 ± 26.51	86.25 ± 29.06	87.25 ± 29.35
Writer-dependent			
PSTM	75.39 ± 26.52	87.88 ± 29.50	88.71 ± 29.75
STM	75.54 ± 26.64	87.25 ± 29.34	88.20 ± 29.61

Table 7.10 Multi-class percentage accuracy comparisons on optdigits dataset

	Training ratio		
	0.01	0.3	0.7
Writer-independent			
PSTM	59.97 ± 2.71	88.12 ± 0.31	89.55 ± 0.64
STM	59.07 ± 2.44	83.58 ± 1.36	87.75 ± 0.98
Writer-dependent			
PSTM	61.46 ± 2.23	92.10 ± 1.00	94.34 ± 0.55
STM	60.63 ± 2.51	86.56 ± 1.39	91.31 ± 0.75

classification, the results of which are summarized in Table 7.10. The average training time comparisons between PSTM and STM, for learning a single binary classifier, are provided in Table 7.11.

Table 7.11 Training time comparisons and number of iterations on optdigits dataset

	Training ratio		
	0.01	0.3	0.7
PSTM	0.005 ± 0.007	0.01 ± 0.060	0.73 ± 0.485
STM	0.02 ± 0.009	0.19 ± 0.067	1.38 ± 0.852
Number of iterations	10 ± 7	14 ± 10	26 ± 24

From the results we can conclude that the efficacy of proximal STM is in line with STM and in terms of time it is much faster. Further, as we increase the size of training dataset we get significant improvement in the testing accuracy.

Remark 7.4.1 Some authors, e.g. Meng et al. [36] have also studied Least Squares Support Tensor Machines (LSSTM). Here it may be be noted that the formulation of LSSTM is very similar to PSTM presented here. The only difference being that in LSSTM, the extra term of $\frac{1}{2}b^2$ present in the objective function of PSTM, is not included. Since we have a detailed discussion of PSTM we omit the details of LSSTM.

7.5 Twin and Least Squares Twin Tensor Machines for Classification

Twin Support Tensor Machine (TWSTM) formulation is a tensor space analogue of TWSVM, which has been proposed by Zhang et al. [37]. In a similar manner, Zhao et al. [38] and Gao et al. [6] introduced its least square version and termed the same as (LS-TWSTM). LS-TWSTM is again a tensor space analogue of the vector space model LS-TWSVM.

Let the given training set be presented by

$$T_C = \{(X_i, y_i), i = 1, 2, \ldots, p\}, \tag{7.58}$$

where $X_i \in \mathbb{R}^{n_1 \times n_2}$ represents the second order tensor and $y_i \in \{-1, 1\}$ are class labels. Thus T_C is essentially a matrix pattern dataset and y_i is the associated class label for the matrix input X_i for $i = 1, 2, \ldots, p$. Let p_1 and p_2 be the number of samples of positive and negative classes respectively, and $p = p_1 + p_2$. Let I_1 and I_2 respectively be the index sets of positive and negative classes, and $|I_1| = p_1, |I_2| = p_2$.

Linear TWSTM finds a pair of nonparallel hyperplanes $f_1(x) = u_1^T X v_1 + b_1 = 0$ and $f_2(x) = u_2^T X v_2 + b_2 = 0$ by constructing two appropriate quadratic programming problems (QPP's). These two QPP's are motivated by (TWSVM1) and (TWSVM2) introduced in Chap. 3 for determining the nonparallel hyperplanes $x^T w_1 + b_1 = 0$ and $x^T w_2 + b_2 = 0$. Thus the two QPP's are

$$(TWSTM1) \qquad \underset{(u_1, v_1, b_1, \xi_2)}{\text{Min}} \quad \frac{1}{2} \sum_{i \in I_1} (u_1 X_i v_1 + b_1)^2 + C_1 \sum_{j \in I_2} \xi_{2j}$$

subject to

$$-(u_1^T X_j v_1 + b_1) + \xi_{2j} \geq 1, \quad j \in I_2,$$
$$\xi_{2j} \geq 0, \quad j \in I_2,$$

and

$$(TWSTM2) \qquad \underset{(u_2, v_2, b_2, \xi_1)}{\text{Min}} \quad \frac{1}{2} \sum_{j \in I_2} (u_2 X_j v_2 + b_2)^2 + C_2 \sum_{i \in I_1} \xi_{1i}$$

subject to

$$(u_2^T X_i v_2 + b_2) + \xi_{1i} \geq 1, \quad i \in I_1,$$
$$\xi_{1i} \geq 0, \quad i \in I_1.$$

Here $u_1, u_2 \in \mathbb{R}^{n_1}$, $v_1, v_2 \in \mathbb{R}^{n_2}$ and b_1, b_2 are decision variables. Further $C_1, C_2 > 0$ are parameters and $\xi_1 \in \mathbb{R}^{p_1}, \xi_2 \in \mathbb{R}^{p_2}$ are slack variables. The geometric interpretation of (TWSTM1) and (TWSTM2) is same as that of (TWSVM1) and (TWSVM2).

Next, as before, a new matrix input X_{new} is assigned the class k depending upon which of the two hyperplanes it is closer to. Thus,

$$\text{label}(X_{new}) = \arg \underset{k=1,2}{Min} \left(\frac{|f_k(X_{new})|}{\|u_k^T v_k\|} \right), \tag{7.59}$$

where $k = 1$ denotes the positive class and $k = 2$ denotes the negative class.

To solve the two quadratic programming problems (TWSTM1) and (TWSTM2) we proceed in the same manner as in the alternating algorithm for (STM1) and (STM2) discussed in Sect. 7.2. We present here the details for (TWSTM1) and the details for (TWSTM2) are analogous.

Algorithm for (TWSTM1)

Step 1 (Initialization)
Choose $\epsilon > 0$ and $u_1^{(0)} = (1, 1, \ldots, 1)^T \in \mathbb{R}^{n_1}$. Set $s = 0$.

Step 2 (Determination of v_1 and b_1)
Given $u_1^{(s)}$ and hence $X_i^T u_1^{(s)}$, $i \in I_1$, solve (TWSTM1) and obtain its optimal solution $(v_1^{(s)}, b_1^{(s)})$.

Step 3 (Updation of u)
Given $v_1^{(s)}$ and hence $X_i v_1^{(s)}$, $i \in I_1$, solve (TWSTM1) and obtain its optimal solution $(u_1^{(s+1)}, b_1^{(s+1)})$.

Step 4 (Updation of v)
Given $u_1^{(s+1)}$ and hence $X_i^T u_1^{(s+1)}$, $i \in I_1$, solve (TWSTM1) and obtain its optimal solution $(v_1^{(s+1)}, \hat{b}_1^{(s+1)})$.

Step 5 (Stopping rule)
If $\|u_1^{(s+1)} - u_1^{(s)}\|$ and $\|v_1^{(s+1)} - v_1^{(s)}\|$ are all less than ϵ or the maximum number of iterations is achieved, stop and take $\overline{u_1} = u_1^{(s+1)}$, $\overline{v_1} = v_1^{(s+1)}$, $\overline{b_1} = \frac{b_1^{(s+1)} + \hat{b}_1^{(s+1)}}{2}$, $f_1(x) = \overline{u_1}^T x \overline{v_1} + \overline{b_1}$. Otherwise put $s \leftarrow s + 1$ and return to Step 2.

The solution of (TWSTM2) gives the second plane $f_2(x) = \overline{u_2}^T x \overline{v_2} + \overline{b_2}$, and then for a new input X_{new}, its class label is assigned as per rule (7.59).

Taking motivation from LS-TWSVM, we can formally write the formulation of LS-TWSTM as follows

$$(LS - TWSTM1) \qquad \underset{(u_1, v_1, b_1, \xi_2)}{\text{Min}} \quad \frac{1}{2} \sum_{i \in I_1} (u_1 X_i v_1 + b_1)^2 + C_1 \sum_{j \in I_2} \xi_{2j}^2$$

subject to

$$-(u_1^T X_j v_1 + b_1) + \xi_{2j} = 1, \quad j \in I_2,$$

and

$$(LS - TWSTM2) \qquad \underset{(u_2, v_2, b_2, \xi_1)}{\text{Min}} \quad \frac{1}{2} \sum_{j \in I_2} (u_2 X_j v_2 + b_2)^2 + C_2 \sum_{i \in I_1} \xi_{1i}^2$$

subject to

$$(u_2^T X_i v_2 + b_2) + \xi_{1i} \geq 1, \quad i \in I_1.$$

Zhao et al. [5] and Gao et al. [6] derived alternating algorithm for (LS-TWSTM) which is similar to (TWSTM) except that QPP (LS-TWSTM1) and (LS-TWSTM2) are solved by solving a system of linear equations. This is the common approach for solving all least squares versions, be it LS-SVM, LS-TWSVM, LS-STM or LS-TWSTM. In fact, one can also consider proximal version of TSTM on the lines of PSTM and can achieve all associated simplifications in the computations.

Zhao et al. [5] and Gao et al. [6] did extensive implementations on Eyes and ORL database and showed that LS-TWSTM performs uniformly better than linear STM and linear TWSTM. Gao et al. [6] also proposed kernel version of LS-TWSTM and termed it NLS-TWSTM. On ORL database, NLS-TWSTM is reported to have better performance than LS-TWSTM. Similar experience has also been reported on Yale database.

Remark 7.5.1 Taking motivation from (GEPSVM), Zhang and Chow [39] developed Maximum Margin Multisurface Proximal Support Tensor Machines (M^{3}PSTM) model and discussed its application to image classification and segmentation. Further, Shi et al. [40] proposed a Tensor Distance Based Least Square Twin Support Tensor Machine (TDLS-TSTM) model which can take full advantage of the structural information of data. Various other extensions of (TWSTM), which are similar to the extensions of (TWSVM), have been reported in the literature.

Remark 7.5.2 In support tensor and twin support tensor type machines, mostly second order tensors are used as the input. Therefore, some authors have also called such learning machines as Support Matrix Machines, e.g. model of Xu et al. [41] for 2-dimensional image data classification. Some other references with this terminology are Matrix Pattern Based Projection Twin Support Vector Machin es (Hua and Ding [42]), New Least Squares Support Vector Machines Based on Matrix Patterns (Wang and Chen [43]) and Multiple Rank kernel Support Matrix Machine for Matrix Data Classification (Gao et al. [32]).

7.6 Conclusions

This chapter studies certain additional topics related with TWSVM. Just as in SVM, the problem of optimal kernel selection in TWSVM is also very important. Here taking motivation from Fung et al. [13] and utilizing the iterative alternating optimization algorithm of Bezdek and Hathaway [12], a methodology is developed for selecting optimal kernel in TWSVM. Classification with prior knowledge in the context of TWSVM is also been discussed in this chapter. We have presented its least squares version, i.e. KBTWSVM and KBLTWSVM, in detail and shown its superiority with other knowledge based formulations.

The last topic discussed in this chapter is Twin Support Tensor Machines. Support tensor machines are conceptually different from SVMs as these are tensor space model and not vector space models. They have found much favour in the areas of text characterization and image processing. This topic seems to be very promising for future studies.

References

1. Khemchandani, R., & Jayadeva Chandra, S. (2009). Optimal kernel selection in twin support vector machines. *Optimization Letters*, *3*, 77–88.
2. Khemchandani, R., Karpatne, A., & Chandra, S. (2013). Proximal support tensor machines. *International Journal of Machine Learning and Cybernetics*, *4*, 703–712.
3. Kumar, M. A., Khemchandani, R., Gopal, M., & Chandra, S. (2010). Knowledge based least squares twin support vector machines. *Information Sciences*, *180*, 4606–4618.
4. Cai, D., He, X., Wen, J.-R., Han, J. & Ma, W.-Y. (2006). Support tensor machines for text categorization. Technical Report, Department of Computer Science, UIUC. UIUCDCS-R-2006-2714.

5. Zhao, X., Shi, H., Meng, L., & Jing, L. (2014). Least squares twin support tensor machines for classification. *Journal of Information and Computational Science, 11–12*, 4175–4189.
6. Gao, X., Fan, L., & Xu, H. (2014). NLS-TWSTM: A novel and fast nonlinear image classification method. *WSEAS Transactions on Mathematics, 13*, 626–635.
7. Lanckriet, G. R. G., Cristianini, N., Bartlett, P., Ghaoui, L. El, & Jordan, M. I. (2004). Learning the kernel matrix with semidefinite programming. *Journal of Machine Learning Research, 5*, 27–72.
8. Xiong, H., Swamy, M. N. S., & Ahmad, M. O. (2005). Optimizing the kernel in the empirical feature space. *IEEE Transactions on Neural Networks, 16*(2), 460–474.
9. Jayadeva Shah, S., & Chandra, S. (2009). Kernel optimization using a generalized eigenvalue approach. *PReMI*, 32–37.
10. Fung, G., Dundar, M., Bi, J., & Rao, B. (2004). A fast iterative algorithm for Fisher discriminant using heterogeneous kernels. In: *Proceedings of the 21st International conference on Machine Learning (ICML)* (pp. 264–272). Canada: ACM.
11. Mika, S., Rätsch, G., & Müller, K. (2001). A mathematical programming approach to the kernel fisher algorithm. In. *Advances in Neural Information Processing Systems, 13*, 591–597.
12. Bezdek, J. C., & Hathaway, R. J. (2003). Convergence of alternating optimization. *Neural, Parallel and Scientific Computations, 11*(4), 351–368.
13. Fung, G., & Mangasarian, O. L. (2001). Proximal support vector machine classifiers. In: F. Provost, & R. Srikant (Eds.), *Proceedings of Seventh International Conference on Knowledge Discovery and Data Mining* (pp. 77–86).
14. Blake, C.L., & Merz, C.J., UCI Repository for Machine Learning Databases, Irvine, CA: University of California, Department of Information and Computer Sciences. http://www.ics.uci.edu/~mlearn/MLRepository.html.
15. Duda, R., Hart, P., & Stork, D. (2001). *Pattern classification*. New York: Wiley.
16. Fung, G., Mangasarian, O. L., & Shavlik, J. (2002). Knowledge- based support vector machine classifiers. In: *Advances in Neural Information Processing Systems* (vol. 14, pp. 01–09). Massachusetts: MIT Press.
17. Mangasarian, O. L. (1994). Nonlinear programming. SIAM.
18. Khemchandani, R., & Jayadeva, Chandra, S. (2009). Knowledge based proximal support vector machines. *European Journal of Operational Research, 195*(3), 914–923.
19. Golub, G. H., & Van Loan, C. F. (1996). *Matrix computations* (3rd ed.). Baltimore, Maryland: The John Hopkins University Press.
20. Jayadeva, Khemchandani., R., & Chandra, S. (2007). Twin support vector machines for pattern classification. *IEEE Transactions on Pattern Analysis and Machine Intelligence, 29*(5), 905–910.
21. Lee, Y. j., Mangasarian, O. L., & Wolberg, W. H. (2003). Survival-time classification of breast cancer patients. *Computational optimization and Applications, 25*, 151–166.
22. Cristianini, N., & Shawe-Taylor, J. (2000). *An introduction to support vector machines and other Kernel-based learning methods*. New York: Cambridge University Press.
23. Vapnik, V. (1998). *Statistical learning theory*. New York: Wiley.
24. Wang, Z., Chen, S. C., liu, J., Zhang, D. Q. (2008). Pattern representation in feature extraction and classifier design: Matrix versus vector. *IEEE Transactions on Neural Networks, 19*, 758–769.
25. Zhang, Z., & Ye, N. (2011). Learning a tensor subspace for semi-supervised dimensionality reduction. *Soft Computing, 15*(2), 383–395.
26. Fu, Y., & Huang, T. S. (2008). Image classification using correlation tensor analysis. *IEEE Transactions on Image Processing, 17*(2), 226–234.
27. He, X., Cai, D., & Niyogi, P. (2005). *Tensor subspace analysis, advances in neural information processing systems*. Canada: Vancouver.
28. Yan, S., Xu, D., Yang, Q., Zhang, L., Tang, X., & Zhang, H. (2007). Multilinear discriminant analysis for face recognition. *IEEE Transactions on Image Processing, 16*(1), 212–220.

29. Tao, D., Li, X., Hu, W., Maybank, S. J., & Wu, X. (2007). General tensor discriminant analysis and Gabor features for gait recognition. *IEEE Transactions on Pattern Analysis and Machine Intelligence*, *29*(10), 1700–1715.
30. Tao, D., Li, X., Hu, W., Maybank, S. J., & Wu, X. (2008). Tensor rank one discriminant analysis: a convergent method for discriminative multilinear subspace selection. *Neurocomputing*, *71*(10–12), 1866–1882.
31. Chandra, S., Jayadeva, Mehra, A. (2009). *Numerical optimization with applications*. New Delhi: Narosa Publishing House.
32. Gao, X., Fan, L., & Xu, H. (2013). Multiple rank kernel support matrix machine for matrix data classification. *International Journal of Machine Learning and Cybernetics*, *4*, 703–712.
33. Hu, R. X., Jia, W., Huang, D. S., & Lei, Y. K. (2010). Maximum margin criterion with tensor representation. *Neurocomputing*, *73*(10–12), 1541–1549.
34. Plamondon, R., & Srihari, S. N. (2000). On-line and off-line handwriting recognition: A comprehensive survey. *IEEE Transactions on Pattern Analysis and Machine Intelligence*, *22*(1), 63–84.
35. Alpaydin, E., & Kaynak, C. (1998). UCI Machine Learning Repository, Irvine, CA: University of California, Department of Information and Computer Sciences. http://archive.ics.uci.edu/ml.
36. Meng, L.V., Zhao, X., Song, L., Shi, H., & Jing, L. (2013). Least squares support tensor machine. ISORA 978-1-84919-713-7, IET.
37. Zhang, X., Gao, X., & Wang, Y. (2009). Twin support tensor machines for MCS detection. *Journal of Electronics*, *26*, 318–324.
38. Zhang, X., Gao, X., & Wang, Y. (2014). Least squares twin support tensor machine for classification. *Journal of Information and Computational Science*, *11–12*, 4175–4189.
39. Zhang, Z., & Chow, T. W. S. (2012). Maximum margin multisurface support tensor machines with application to image classification and segmentation. *Expert Systems with Applications*, *39*, 849–860.
40. Shi, H., Zhao, X., & Jing, L. (2014). Tensor distance based least square twin support tensor machine. *Applied Mechanics and Materials*, *668–669*, 1170–1173.
41. Xu, H., Fan, L., & Gao, X. (2015). Projection twin SMM's for 2d image data classification. *Neural Computing and Applications*, *26*, 91–100.
42. Hua, X., & Ding, S. (2012). Matrix pattern based projection twin support vector machines. *International Journal of Digital Content Technology and its Applications*, *6*(20), 172–181.
43. Wang, Z., & Chen, S. (2007). New least squares support vector machines based on matrix patterns. *Neural Processing Letters*, *26*, 41–56.

Chapter 8
Applications Based on TWSVM

8.1 Introduction

The Twin Support Vector Machine (TWSVM) has been widely used in a variety of applications based on classification (binary, multi-class, multi-label), regression (function estimation) and clustering tasks. The key benefit of using the TWSVM as an alternative to the conventional SVM arises from the fact that it is inherently designed to handle the class imbalance problem and is four times faster than SVM, which makes it usable for real-time applications. Such a situation is common in various practical applications for classification and regression, particularly in the case of multi-class classification. This makes the TWSVM a natural choice for such applications, resulting in better generalization (higher accuracy) as well as in obtaining a faster solution (lower training time as a result of solving smaller sized Quadratic Programming Problems - QPPs).

In order to deal with unbalanced dataset some of the recent techniques includes randomly oversampling the smaller class, or undersampling the larger class; one sided selection - which involves removing examples of the larger class that are "noisy" [1], and cluster-based oversampling - which eliminates small subsets of isolated examples [2]. Wilson's editing [3] employs a 3-nearest neighbour classifier to remove majority class examples SS that are misclassified. SMOTE [4] adds minority class exemplars by extrapolating between existing ones. A summary of many techniques and comparisons may be found in [5, 6]. All these techniques modify the training dataset in some manner. In contrast, the TWSVM does not change the dataset at all.

This chapter aims to review a few significant applications which have benefited from using the TWSVM. The objective is to illustrate the widespread applicability of the TWSVM across a multitude of application domains, while enabling the reader to gain insight to identify potential application domains. Research papers cited in this chapter are indexed in online search forums like Google Scholar and Scopus.

Jayadeva et al., *Twin Support Vector Machines*, Studies in Computational Intelligence 659, DOI 10.1007/978-3-319-46186-1_8

8.2 Applications to Bio-Signal Processing

Bio-signal processing has been an interesting application benefited by the use of the TWSVM. It is naturally suited for classification of such data owing to the inherent nature of class imbalance present in many such problems. These include applications based on bio signals such as Electroencephalogram (brain signals), Electromyogram (muscle activity), Electrocardiogram (heart beat), Electrooculogram (eye movements), Galvanic Skin Response and Magnetoencephalogram, or even a combination of these as in the case of multi-modal systems. Such systems usually require classification of intents generated by users through such signals in order to initiate a command or an action. One can understand that as this would require to classify several intents, apart from a *neutral* state which would correspond to the resting state of the user. This is precisely the multi-class classification problem, where the classifier would need to detect one intent among a set of intents, and the TWSVM is suited to such scenarios. We discuss two specific examples of biosignals where the TWSVM has demonstrated utility in the following sub-sections.

Consider for instance an application which involves building a machine learning pipeline for detecting facial gestures from muscular activity detected from Electromyogram (EMG) signals. There may be N facial movements which need to be classified, and this would usually be achieved by a one-v/s-rest scheme for classifiers. Using a conventional classifier such as the SVM for such a purpose results in building classifiers biased towards the class with larger number of samples. As TWSVM's formulation is designed to tackle this problem, the use of TWSVM in such scenarios leads to improved generalization. This is particularly significant considering that the number of samples available for such scenarios is low, as acquiring a large number of training samples is often inconvenient for the user. Thus, making each classifier robust becomes essential. In addition, it also offers faster classifier training to some extent as smaller sized problems are solved.

8.2.1 Surface Electromyogram (sEMG) Data

The TWSVM has been successfully used in gesture classification using Surface Electromyogram (sEMG) signals. The problem at hand in this context is to train classifiers using sEMG data corresponding to limb movements such as wrist and finger flexions, such as those shown in Fig. 8.1. The data is collected by attaching electrodes to the limb that pick up nerve impulses when the subject performs limb movements. These electrodes, usually non-invasive, are called channels in biomedical signal parlance.

sEMG datasets are high-dimensional and unbalanced. The high dimensionality comes from the fact that these electrodes typically acquire data at a sampling rate, of say S samples per second, and when this data is collected from k electrodes (channels) for a time period of t seconds, the size of the data matrix is of the order of $S \times k \times t$.

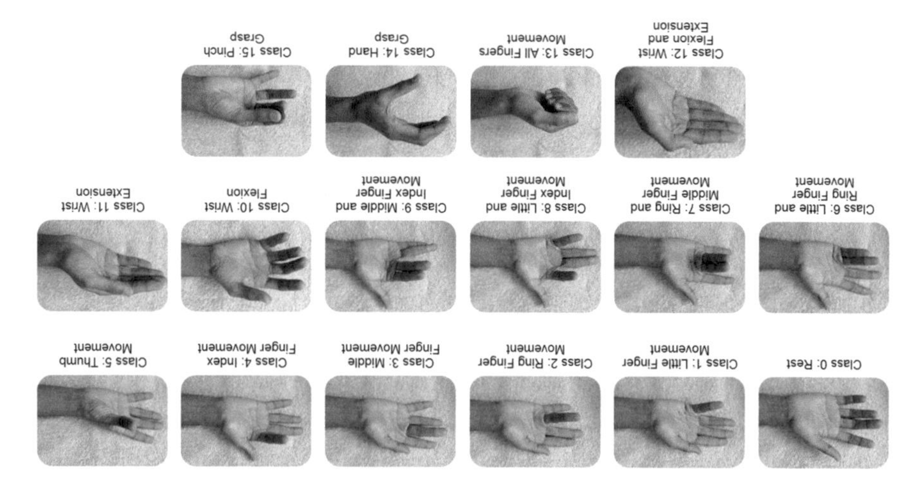

Fig. 8.1 Various wrist and finger flexions that are classified from sEMG data in [7, 8]

For example, a recording of a finger flexion for 5 s from 4 channels sampled at 250 samples per second would have 5000 samples.

The second challenge in building efficient classifiers is the class imbalance problem inherent to such datasets. This is essentially due to two restrictions, first that collecting a large number of samples is tedious for the user, and secondly, being a multi-class problem, the use of one-v/s-rest classification scheme results in a large number of samples for the "other" class samples. To give an example, suppose we are trying to classify $I_1, I_2, \ldots, I_p$ intents (wrist and finger flexions), and we have k samples of each of the p intents, then each classifier in the one-v/s-rest scheme would have k samples of an intent $I_i, i \in \{1, p\}$ and $(k-1) \times p$ samples corresponding to "other" intents $I_j, j \in \{1, p\}, j \neq i$. This creates class imbalance, which amplifies as the number of intents to be detected increases. Thus, a classifier that is insensitive to class imbalance finds suitable applications in this domain.

Naik et al. [7] show that the use of the TWSVM results in a sensitivity and specificity of 84.8 and 88.1 %, respectively, and opposed to neural network sensitivity and specificity being 58.3 and 42.5 %, respectively for the same set of gestures using Root Mean Square (RMS) values as features. Datasets corresponding to seven gestures (various wrist and finger flexions) were investigated for seven subjects in this study. Similarly, Arjunan et al. [9] used fractal features on similar data and obtained recognition accuracies ranging from 82 to 95 % for the Radial Basis Function (RBF) kernel, and 78 to 91 % for linear kernel TWSVM. Further, promising results on amputee subjects have been reported by Kumar et al. [8], where they obtain accuracy and sensitivity of identification of finger actions from single channel sEMG signal to be 93 and 94 % for able-bodied, and 81 and 84 % for trans-radial amputated respectively. More recently, Arjunan et al. [10] have shown significant improvement in grip recognition accuracy, sensitivity, and specificity obtaining kappa coefficient $\kappa = 0.91$, which is indicative of the TWSVM's robustness. These results are significant in the development of neuro-prosthetic devices for emulating limb movements in disabled subjects, as well as other commercial applications based on assistive/augmented technology.

8.2.2 Electroencephalogram (EEG) Data

The TWSVM has also been shown to be beneficial for classification of Electroencephalogram (EEG) data by Soman and Jayadeva [11] Their work investigates the use of EEG data corresponding to imagined motor movements that are useful for developing Brain Computer Interface (BCI) systems. These use voluntary variations in brain activity to control external devices, which are useful for patients suffering from locomotor disabilities. In practice, these systems (the specific class of motor-imagery BCI systems) allow disabled people to control external devices by imagining movements of their limbs, which are detected from the EEG data. Applications include movement of a motorized wheelchair, movement of the mouse cursor on a computer screen, synthesized speech generation etc.

This problem faces challenges similar to the challenges with sEMG data, and this has been addressed in [11] by developing a processing pipeline using the TWSVM and a feature selection scheme that reduces computation time and leads to higher accuracies on publicly available benchmark datasets. Conventionally, the EEG signals are high dimensional, as is the case with sEMG. In order to work in a space of reasonable dimensionality, the EEG signals are projected to a lower dimensional space [12], and classification is done in this feature space. This projection presents an inherent challenge in terms of loss of information, specifically as a trade-off between being able to retain either time or frequency information. Later methods improved upon this by considering the dimensionality reduction after filtering the EEG data in multiple frequency bands [13, 14]. However, this led to an increase in the number of features and an efficient selection scheme was needed. In [11], a computationally

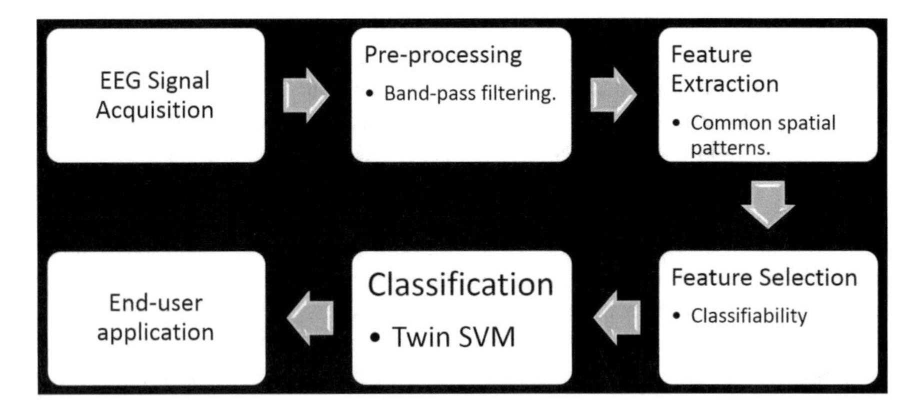

Fig. 8.2 The EEG processing pipeline using TWSVM for classification

simple measure called the classifiability [15] has been used to accomplish this. The summary of the processing pipeline employed is shown in Fig. 8.2.

Their results indicate a relative improvement on classification accuracy ranging from 6.60 % to upto 43.22 % against existing methods. On multi-class datasets, the use of the TWSVM over the conventional SVM clearly results in an improvement in the classification accuracy (in the range of 73–79% as opposed to SVM accuracies of only upto 49 %).

8.3 Applications in the Medical Domain

Other than the applications pertaining to biomedical signal processing, the TWSVM has also been successfully used to build systems to supplement medical disease diagnosis and detection. A few such examples are discussed in this section. In the context of Micro-Calcification Cluster (MCC) detection, TWSVM has been shown to be useful for detecting clustered micro-calcifications in mammogram images. This is useful for early detection of breast cancer, and hence finds importance in from the clinical perspective. The datasets are from the Digital Database for Screening Mammography (DDSM), which has been annotated for ground truth by radiologists. Various combinations of features are used to build classifiers, which include intensity histograms, statistical moments, texture-based features and transform domain features such as wavelet transform coefficients. Zhang et al. [16, 17] use a boosted variant of the TWSVM resulting in a sensitivity of 92.35 % with an error rate of 8.3 %. Further, the tensor variant of the TWSVM has been explored in [18], and results indicate it addresses the problem of over-fitting faced previously. The tensor variant results in a sensitivity and specificity of 0.9385 and 0.9269 respectively, and an average error rate of 0.0693, outperforming the SVM.

Si and Jing [19] have developed Computer-Aided Diagnostic (CAD) system for mass detection in digital mammograms. Their results are based on the Digital Database for Screening Mammography (DDSM) dataset, and reveal that 94 % sensitivity was achieved for malignant masses and 78 % detection for benign masses.

Disease detection systems have been developed using the TWSVM such as for Parkinsons by Tomar et al. [20], which uses the LS-TWSVM in conjunction with Particle Swarm Optimization (PSO) to obtain an accuracy of 97.95 %, that is superior to 11 other methods reported in literature [20, Table 1]. These authors have also developed a system for detection of heart disease using the TWSVM [21] and obtain an accuracy of 85.59 % on the heart-statlog dataset. These results indicate the feasibility of using the TWSVM for problems pertaining to disease detection and diagnosis systems.

8.4 Applications to Intrusion and Fault Detection Problems

Problems such as intrusion and fault detection are fall into the category of applications that benefit by use of classifiers such as the TWSVM. Intrusion detection is of significance in wireless networks as it is aids development of efficient network security protocols. Examples of intrusions to be detected include Denial of Service attack, user to root attack, remote to local user attack and probing. The challenge here is to be able to build a classifier that detects intrusions effectively, and this is non-trivial as the training data available would have very few number of labeled intrusions as these are comparatively infrequent in terms of occurrence over time. This results in the class imbalance problem. The case is similar for detection of faults in mechanical systems. Machine learning techniques have been applied in literature to automate the detection of intrusions or faults, and using the Twin SVM has shown a significant improvement in the detection rate in comparison.

The TWSVM has been used for intrusion detection by Ding et al. [22] on the KDD 1999 dataset and obtain higher detection rates and lower training time. The objective was to build a predictive model capable of distinguishing between "bad" connections, called intrusions or attacks, and "good" normal connections. They have used basic features (derived from packet headers), content features (based on domain knowledge of network packets) time-based and host-based features. The results show improvement in detection rates across various categories of network intrusions considered, for both the original set of 41 features as well as on a reduced feature set. The results have been compared with the Wenke-Lee method and the conventional SVM. For the case of using all features, there is an improvement of 0.75 % over the traditional SVM model, and 19.06 % than the Wenke-Lee model. For the reduced feature set, there is an improvement of 0.59 % than the SVM model and 19.48 % over the Wenke-Lee model. There is also significant reduction in training time as opposed to conventional SVMs, of upto 75 % for the full feature set and 69 % on the reduced feature set. Further, a probability based approach by Nei et al. [23] using a network flow feature selection procedure obtains upto 97.94 % detection rate on intrusion detection on the same dataset.

Another application of detecting defects in computer software programs using the TWSVM has been investigated by Agarwal et al. [24]. They have used the CM1 PROMISE software engineering repository dataset which contains the information such as Lines of Code (LOC), Design Complexity, Cyclomatic Complexity, Essential Complexity, Effort Measures, comments and various other attributes that are useful for predicting whether a software has defects or not. They obtain an accuracy of 99.10 % using the TWSVM while 12 other state-of-the-art methods obtain accuracies in the range of 76.83–93.17 %.

Ding. et al. [25] have devised a diagnosis scheme for fault detection in wireless sensors based on using the Twin SVM, using Particle Swarm Optimization (PSO) for its parameter tuning. Four types of faults, shock, biasing, short circuit, and shifting were investigated and the performance of TWSVM was compared with other diagnostic methods, which used neural networks and conventional SVMs. These methods suffered from the drawbacks such as getting stuck in local minima and overfitting for neural networks. Their results indicate that the diagnosis results for wireless sensor using TWSVMs are better than those of SVM and Artificial Neural Network (ANN) - 96.7 % accuracy as opposed to 91.1 % (SVM) and 83.3 %(ANN).

Chu et al. [26] have investigated the use of a variant of the TWSVM, called the multi-density TWSVM (MDTWSVM) for strip steel surface defect classification. This problem aims to identify defects such as scarring, dents, scratches, dirt, holes, damage and edge-cracks on steel surface, and is essentially a multi-class classification problem. They use geometry, gray level, projection, texture and frequency-domain features for classification. The authors have proposed the MDTWSVM for large-scale, sparse, unbalanced and corrupted training datasets. Their formulation incorporates density information using Kernel Density Estimation (KDE) to develop the MDTWSVM for this specific application context. The formulation is a derivative of the Twin SVM formulation in principle. The results report an accuracy of 92.31 % for the SVM, 93.85 % for the TWSVM and 96.92 % for the MDTWSVM. The training times are 91.1512, 11.3707 and 1.2623 s for the SVM, TWSVM and MDTWSVM respectively.

Similarly, Shen et al. [27] show application of the TWSVM for fault diagnosis in rolling bearings. Here too, the TWSVM is useful as the number of faults constitute very few of the total number of samples. They obtain higher accuracies and lower training time using TWSVM as over conventional SVM. These results show significant improvement by the use of TWSVM and similarly motivated formulations.

8.5 Application to Activity Recognition

Human Activity Recognition is an active area of research in Computer Vision. In the last decade a large amount of research has been done in this area, nevertheless, it is still an open and challenging problem. It involves automatically detecting, tracking and recognizing human activities from the information acquired through sensors eg. a sequence of images captured by video cameras. The research in this area is motivated by many applications in surveillance systems, video annotations and human-computer interactions.

A human activity recognition problem is generally approached in two phases: feature extraction followed by classification and labelling of activities. In the classification phase, many methods have been applied in the field of human action recognition, including the k-Nearest Neighbor, Support Vector Machine (SVM), boosting-based classifiers and Hidden Markov Models.

The recent research has witnessed the evolution of support vector machines as a powerful paradigm for pattern classification. But in the application of action recognition, SVM has some drawbacks such as high computational time, sensitivity to noise and outliers; and the unbalanced dataset problem. Owing to these drawbacks, the problem is naturally suited to TWSVM based classifiers which can handle these problem in comparatively robust manner. Since LSTWSVM is comparatively faster than SVM because it solves system of linear equations instead of solving two QPPs to obtain the non-parallel hyperplanes, it has gained popularity in action classification and labelling task [28–30]. Mozafari et al. [28] introduce the application of LSTWSVM in activity recognition system with the local space-time features. In order to handle the intrinsic noise and outliers present in the related activity classes, Nasiri et al. [30] proposed Energy-based least squares twin support vector machine (ELS-TWSVM) on the lines of LSTWSVM for activity recognition in which they changed the unit distance constraints used in LSTWSVM by some pre-defined energy parameter. Similar to LSTWSVM, the ELS-TWSVM [30] seeks a pair of hyperplane given as follows:

$$w_1^T x + b_1 = 0 \text{ and } w_2^T x + b_2 = 0 \tag{8.1}$$

The primal problem of ELS-TWSVM is given below

$$\begin{array}{ll} (\text{ELS-TWSVM 1}) & \underset{w_1, b_1, y_2}{Min} \quad \frac{1}{2}||Aw_1 + e_1 b_1)||^2 + \frac{c_1}{2} y_2^T y_2 \\ & \text{subject to } -(Bw_1 + e_2 b_1) + y_2 = E_1, \end{array} \tag{8.2}$$

and

$$\begin{array}{ll} (\text{ELS-TWSVM 2}) & \underset{w_2, b_2, y_1}{Min} \quad \frac{1}{2}||Bw_2 + e_2 b_2||^2 + \frac{c_2}{2} y_1^T y_1 \\ & \text{subject to } (Aw_2 + e_1 b_2) + y_1 = E_2, \end{array} \tag{8.3}$$

where E_1 and E_2 are user defined energy parameters.

On the similar lines of LSTWSVM, the solution of QPP (8.2) and (8.3) is obtained as follows

$$[w_1 \;\; b_1]^T = -[c_1 G^T G + H^T H]^{-1}[c_1 G^T E_1] \tag{8.4}$$

and

$$[w_2 \;\; b_2]^T = [c_2 H^T H + G^T G]^{-1}[c_2 H^T E_2] \tag{8.5}$$

where $H = [A \quad e_1]$ and $G = [B \quad e_2]$.

A new data point $x \in R^n$ is assigned to class i ($i = +1$ or -1) using the following decision function: $f(x) = \begin{cases} +1, & if \ \frac{|x^T w_1 + e_1 b_1|}{|x^T w_2 + e_2 b_2|} \leq 1, \\ -1, & if \ \frac{|x^T w_1 + e_1 b_1|}{|x^T w_2 + e_2 b_2|} \geq 1. \end{cases}$ ELS-TWSVM takes advantage of prior knowledge available in human activity recognition problem about the uncertainty and intra-class variations and thus improves the performance of the activity recognition to some degree [30].

Various experiments were carried out in [30] on general activity recognition datasets like Weizmann dataset and KTH dataset, which showed the efficacy of the ELS-TWSVM. They have extracted features extraction using Harris detector algorithm and Histogram of Optical Flow descriptors on the KTH database containing six human actions (walking, jogging, running, boxing, hand waving, hand clapping) performed by 25 subjects in various outdoor and indoor environments. This is hence a multi-class problem with high class imbalance, which mandates the use of a TWSVM like formulation. Their findings report that the Least Squares TWSVM (LSTWSVM) obtains 85.56 % accuracy while the ELS-TWSVM obtains 95.56 % accuracy on the Weizmann dataset, while on the KTH dataset, the LSTWSVM gives 92.33 % and the ELS-TWSVM gives 95.21 %. Prior to this, Mozafari [28] demonstrated improvements on the KTS dataset using LSTWSVM and obtained upto 97.2 % accuracy on the dataset.

However, one drawback in ELS-TWSVM is that energy parameter has to be externally fixed which sometimes leads to instability in problem formulations which effects the overall prediction accuracy of the system.

To overcome this limitation and in order to obtain a robust model, Khemchandani et al. [31] introduce a pair of parameters ν and variables ρ_1 and ρ_2 in order to control noise and outliers. Further the optimal values of ρ_1 and ρ_2 are obtained as a part of optimization problem. On the lines of Shao et al. [32], an extra regularization term $\frac{1}{2}w^T w + b^2$ is introduced with the idea of implementing structural risk minimization and also to address the problem of ill-posed solution encountered by ELS-TWSVM. The regularization term ensures the positive definiteness of the optimization problem involved.

In order to incorporate the structural risk minimization principle, Khemchandani et al. [31] propose a robust and efficient classifier based on LS-TWSVM called Robust Least Squares Twin Support Vector Machine (RLS-TWSVM), where this energy parameter is optimized in the problem itself using the available training data. Further, for nonlinear RLS-TWSVM, Shermann-Morrison-Woodbury (SMW) formula and partition method [33] to were used in order to further reduce the training time for learning the classifier.

In RLS-TWSVM the patterns of respective classes are clustered around the following hyperplanes.

$$w_1^T x + b_1 = \rho_1 \text{ and } w_2^T x + b_2 = \rho_2 \tag{8.6}$$

The primal problem of RLS-TWSVM are given as follows

$$\text{(RLS-TWSVM 1)} \quad \underset{w_1, b_1, \rho_1, \xi}{Min} \quad \frac{1}{2}||Aw_1 + e_1 b_1||^2 + \nu\rho_1^T \rho_1 + \frac{c_1}{2}(w_1^T w_1 + b_1^2) + c_2 \xi^T \xi$$
$$\text{subject to} \quad -(Bw_1 + e_2 b_1) = e_2(1 - \rho_1) - \xi, \tag{8.7}$$

where $c_1, c_2 > 0$, e_1 and e_2 are vector of ones of appropriate dimensions; and ξ is the error variable. The parameter $(1 - \rho_1)$ replaced E_1 of ELS-TWSVM, further $\nu > 0$ control the bound on the fractions of support vectors and hence its optimal value ensure the effect of outliers has been taken care of in activity recognition framework. Here, ELS-TWSVM can be thought of as a special case of RLS-TWSVM where the energy parameter is fixed to the optimal value of $(1 - \rho_1)$.

Khemchandani et al. [31] further introduced the hierarchical framework to deal with multi-category classification problem of activity recognition system in which the hierarchy of classes is determined using the using different multi-category classification approaches discussed in Chap. 5. Experimental results carried out in [31] using global space-time features demonstrated that ELS-TWSVM mostly outperforms other state of art methods. RLS-TWSVM obtains 100 % prediction accuracy on Weizmann dataset.

8.6 Applications in Image Processing

Problems involving image processing, specifically problems such as object detection, have been benefited by the use of the TWSVM. These problems also suffer from the malady of class imbalance, for instance when considering a one-v/s-rest scheme for classification of images belonging to N categories, or in other words, detecting the presence/absence of N objects in these set of images. The developments in feature extraction techniques play a significant role in this as well, however the classifier used is indispensable in obtaining a system with high accuracies.

In the context of image classification, a Ternary Decision Structure based multi-category TWSVM (TDS-TWSVM) classifier has been developed by Khemchandani et al. [34] The aim of the work is to facilitate Content-Based Image Retrieval (CBIR). Typically, CBIR uses low-level features like color, texture, shape, and spatial layout along with semantic features for indexing of images. The authors have used Complete Robust Local Binary Pattern(CRLBP) and Angular Radial Transform (ART) features in their study. They obtain a retrieval rate precision of 64.15 % on the COREL 5K dataset, 93.51 % for the MIT VisTex dataset [35], 85.56 % on the Wang's dataset and 74.35 % on the OT-Scene dataset, outperforming several state-of-the-art approaches. The experiments conclude that TDS-TWSVM outperforms One-Against-All TWSVM (OAA-TWSVM) and binary tree-based TWSVM(TB-TWSVM)in terms of classification accuracy. Image de-noising has been investigated by Yang et al. [36] using Non-Subsampled Shearlet Transform (NSST) and the TWSVM, and this has led to improvement in Signal-to-Noise (SNR) ratios in images from various domains, outperforming results of 5 other benchmark approaches.

8.7 Applications to Finance

There have also been applications of the TWSVM in financial predictions and modelling. For instance, Ye et al. [37, 38] have used the TWSVM to identify the determinants in inflation for the Chinese economy by considering financial data from two distinct economic periods. Fang et al. [39] have used the economic development data of Anhui province from 1992 to 2009 to study economic development prediction using wavelet kernel-based primal TWSVM, and have obtained improvement over the conventional SVM.

8.8 Applications for Online Learning

The TWSVM has also been used to devise online learning schemes, which are beneficial for large datasets and streaming data that occurs in many practical applications. These models also adapt the classifier model based on predictions determined on the incoming samples, in cases where feedback on correctness can be provided. Such models find widespread applications in real-time scenarios. For instance, Khemchandani et al. [40] have devised the incremental TWSVM, Hao et al. [41] devised the fast incremental TWSVM and Mu et al. [42] have developed the online TWSVM.

8.9 Other Applications

The TWSVM has been used for achievement analysis of students by Yang and Liu [43], where the use of the twin parametric-margin SVM has been used to determine student's achievement levels "Advanced language program design" course. The performance has been compared against conventional SVM and neural networks. The evaluation results for student achievement based on weighted TWSVM network are consistent with the actual results, and also performs better than the comparative methods investigated. Similar findings have also been reported by Lui et al. [44].

Manifold fuzzy TWSVMs have been investigated by Liu et al. [45] for classification of star spectra. Experiments with alternate classification methods, such as SVM and k-nearest neighbors on the datasets illustrate the superior results obtained using TWSVM. Aldhaifallah et al. [46] have identified Auto-Regressive Exogenous Models (ARX) based on Twin Support Vector Machine Regression (TWSVR). Their results indicate that the use of the TWSVR outperforms SVR and Least Squares SVR (LSSVR) in terms of accuracy. Moreover, the CPU time spent by the TWSVR algorithm is much less than the time spent by SVR algorithm. Other recognition applications include license plate recognition by Gao et al. [47].

8.10 Conclusions

This chapter presented an array of applications and practical problems benefited by the use of the TWSVM and its variants. As it would have become evident to the reader, the TWSVM addresses a significant problem occurring in practical datasets, that of class imbalance, and suitably modifies the conventional SVM formulation to tackle the same. In addition, this results in solving smaller sized QPPs, that results in lower computation time in training classifier models. The TWSVM has been extended to various formulations, and these too have benefited practical applications. Future work lies in evolving TWSVM based formulations for newer learning paradigms such as deep learning and allied hybrid models, which may possibly benefit from the advantages of the TWSVM. Minimal complexity implementations of the TWSVM are of interest as well, as these are relevant for realization on hardware where design area and power consumption are of significance. The galore of practical applications continues to grow by the day with the untamed growth of available data. Hence, the relevance of such algorithms, and machine learning in general, continues to increase.

References

1. Kubat, M., & Matwin, S. (1997). Addressing the curse of imbalanced training sets: One-sided selection. *(ICML)*, *97*, 179–186 (1997).
2. Jo, T., & Japkowicz, N. (2004). Class imbalances versus small disjuncts. *ACM SIGKDD Explorations Newsletter*, *6*(1), 40–49.
3. Barandela, R., Valdovinos, R. M., Sánchez, J. S., & Ferri, F. J. (2004). The imbalanced training sample problem: Under or over sampling. *Structural, Syntactic, and Statistical Pattern Recognition*, 806–814.
4. Chawla, N. V., Bowyer, K. W., Hall, L. O., & Kegelmeyer, W. P. (2002). SMOTE: Synthetic minority over-sampling technique. *Journal of Artificial Intelligence Research*, 321–357.
5. Van H. J., Khoshgoftaar, T. M., & Napolitano, A. (2007). Experimental perspectives on learning from imbalanced data. In *Proceedings of the 24th International Conference on Machine learning* (pp. 935–942).
6. Estabrooks, A., Jo, T., & Japkowicz, N. (2004). A multiple resampling method for learning from imbalanced data sets. *Computational Intelligence*, *20*(1), 18–36.
7. Naik, R., & Kumar, D., & Jayadeva., (2010). Twin SVM for gesture classification using the surface electromyogram. *IEEE Transactions on Information Technology in Biomedicine*, *14*(2), 301–308.
8. Kumar, D. K., Arjunan, S. P., & Singh, V. P. (2013). Towards identification of finger flexions using single channel surface electromyography-able bodied and amputee subjects. *Journal of NeuroEngineering and Rehabilitation*, *10*(50), 1–7.
9. Arjunan, S. P., Kumar, D. K., & Naik, G. R. (2010). A machine learning based method for classification of fractal features of forearm sEMG using twin support vector machines, engineering in medicine and biology society (EMBC). In: *2010 Annual international conference of the IEEE* (pp. 4821–4824).
10. Arjunan, S. P., Kumar, D. K., & Jayadeva. (2015). Fractal and twin SVM-based handgrip recognition for healthy subjects and trans-radial amputees using myoelectric signal, Biomedical Engineering/Biomedizinische Technik. doi:10.1515/bmt-2014-0134.
11. Soman, S., & Jayadeva., (2015). High performance EEG signal classification using classifiability and the Twin SVM. *Applied Soft Computing*, *30*, 305–318.
12. Ramoser, H., Muller-Gerking, J., & Pfurtscheller, G. (2000). Optimal spatial filtering of single trial EEG during imagined hand movement. *IEEE Transactions on Rehabilitation Engineering*, *8*(4), 441–446.
13. Ang, K. K., Chin, Z. Y., Zhang, H., & Guan, C. (2008). Filter bank common spatial pattern (FBCSP) in brain-computer interface. In *IEEE International Joint Conference on Neural Networks, 2008. IJCNN, 2008* (pp. 2390–2397). IEEE World Congress on Computational Intelligence,
14. Novi, Q., Guan, C., Dat, T. H., & Xue, P. (2007). Sub-band common spatial pattern (SBCSP) for brain-computer interface. In *3rd International IEEE/EMBS Conference on Neural Engineering, 2007. CNE'07* (pp. 204–207).
15. Li, Y., Dong, M., & Kothari, R. (2005). Classifiability-based omnivariate decision trees. *IEEE Transactions on Neural Networks*, *16*(6), 1547–1560.

16. Zhang, X. (2009). Boosting twin support vector machine approach for MCs detection. In *Asia-Pacific Conference on Information Processing, 2009. APCIP 2009* (vol. 1, pp. 149–152).
17. Zhang, X., Gao, X., & Wang, Y. (2009). MCs detection with combined image features and twin support vector machines. *Journal of Computers, 4*(3), 215–221.
18. Zhang, X., Gao, X., & Wang, Y. (2009). Twin support tensor machines for MCs detection. *Journal of Electronics (China), 26*(3), 318–325.
19. Si, X., & Jing, L. (2009). Mass detection in digital mammograms using twin support vector machine-based CAD system. In *WASE International Conference on Information Engineering, 2009. ICIE'09* (vol. 1, pp. 240–243). New York: IEEE.
20. Tomar, D., Prasad, B. R., Agarwal, S. (2004). An efficient parkinson disease diagnosis system based on least squares twin support vector machine and particle swarm optimization. In *2014 9th International IEEE Conference on Industrial and Information Systems (ICIIS)* (pp. 1–6).
21. Tomar, D., & Agarwal, S. (2014). Feature selection based least square twin support vector machine for diagnosis of heart disease. *International Journal of Bio-Science and Bio-Technology, 6*(2), 69–82.
22. Ding, X. Zhang, G., Ke, Y., Ma, B., & Li, Z. (2008). High efficient intrusion detection methodology with twin support vector machines. In *International Symposium on Information Science and Engineering, 2008. ISISE'08* (vol. 1, pp. 560–564).
23. Nie, W., & He, D. (2010). A probability approach to anomaly detection with twin support vector machines. *Journal of Shanghai Jiaotong University (Science), 15*, 385–391.
24. Aggarwal, S., Tomar, D., & Verma, S. (2014). Prediction of software defects using twin support vector machine. In *IEEE International Conference on Information Systems and Computer Networks (ISCON)* (pp. 128–132).
25. Ding, M., Yang, D., & Li, X. (2013). Fault diagnosis for wireless sensor by twin support vector machine. In *Mathematical Problems in Engineering* Cairo: Hindawi Publishing Corporation.
26. Chu, M., Gong, R., & Wang, A. (2014). Strip stee surface defect classification method based on enhanced twin support vector machine. *ISIJ International (the Iron and Steel Institute of Japan), 54*(1), 119–124.
27. Shen, Z., Yao, N., Dong, H., & Yao, Y. (2014). Application of twin support vector machine for fault diagnosis of rolling bearing. In *Mechatronics and Automatic Control Systems* (pp. 161–167). Heidelberg: Springer.
28. Mozafari, K., Nasir, J., Charkar, N. M., & Jalili, S. (2011). Action recognition by local space-time features and least square twin SVM (LS-TSVM). *First International IEEE Conference Informatics and Computational Intelligence (ICI)* (pp. 287–292).
29. Mozafari, K., Nasiri, J. A., Charkari, N. M., & Jalili, S. (2011). Hierarchical least square twin support vector machines based framework for human action recognition. In *2011 9th Iranian Conference on Machine Vision and Image Processing (MVIP)* (pp. 1-5). New York: IEEE.
30. Nasiri, J. A., Charkari, N. M., & Mozafari, K. (2014). Energy-based model of least squares twin support vector machines for human action recognition. *Signal Processing, 104*, 248–257.
31. Khemchandani, R., & Sharma, S. (2016). Robust least squares twin support vector machine for human activity recognition. *Applied Soft Computing*.
32. Shao, Y.-H., Zhang, C.-H., Wang, X.-B., & Deng, N.-Y. (2011). Improvements on twin support vector machines. *IEEE Transactions on Neural Networks, 22*(6), 962–968.
33. Golub, G. H., & Van Loan, C. F. (1996). *Matrix computations* (3rd ed.). Baltimore, Maryland: The John Hopkins University Press.
34. Khemchandani, R., & Saigal, P. (2015). Color image classification and retrieval through ternary decision structure based multi-category TWSVM. *Neurocomputing, 165*, 444–455.
35. Pickard, R., Graszyk, C., Mann, S., Wachman, J., Pickard, L., & Campbell, L. (1995). Vistex database. In *Media Lab* Massachusetts: MIT Cambridge.
36. Yang, H. Y., Wang, X. Y., Niu, P. P., & Liu, Y. C. (2014). Image denoising using nonsubsampled shearlet transform and twin support vector machines. *Neural Networks, 57*, 152–165.
37. Ye, Y. F., Liu, X. J., Shao, Y. H., & Li, J. Y. (2013). L1-ε-Twin support vector regression for the determinants of inflation: A comparative study of two periods in China. *Procedia Computer Science, 17*, 514–522.

38. Ye, Y., Shao, Y., & Chen, W. (2013). Comparing inflation forecasts using an ε-wavelet twin support vector regression. *Journal of Information and Computational Science, 10*, 2041–2049.
39. Fang, S., & Hai Yang, S. (2013). A wavelet kernel-based primal twin support vector machine for economic development prediction. In *Mathematical Problems in Engineering* Hindwani Publishing Corporation.
40. Khemchandani, R., Jayadeva, Chandra, S. (2008). Incremental twin support vector machines. In S. K. Neogy, A. K. das, & R. B. Bapat, (Eds.), *International Conference on Modeling, Computation and Optimization. Published in Modeling, Computation and Optimization, ICMCO-08* Singapore: World Scientific.
41. Hao, Y., Zhang, H., (2014). A fast incremental learning algorithm based on twin support vector machine. In *Seventh International IEEE Symposium on Computational Intelligence and Design (ISCID)* (vol. 2, pp. 92–95).
42. Mu, X., Chen, L., & Li, J. (2012). Online learning algorithm for least squares twin support vector machines. *Computer Simulation, 3*, 1–7.
43. Yang, J., & Liu, W. (2014). A structural twin parametric-margin support vector model and its application in students' achievements analysis. *Journal of Computational Information Systems, 10*(6), 2233–2240.
44. Liu, W. (2014). Student achievement analysis based on weighted TSVM network. *Journal of Computational Information Systems, 10*(5), 1877–1883.
45. Liu, Z.-B., Gao, Y.-Y., & Wang, J.-Z. (2015). Automatic classification method of star spectra data based on manifold fuzzy twin support vector machine. *Guang Pu Xue Yu Guang Pu Fen Xi/Spectroscopy and Spectral Analysis, 35*(1), 263–266.
46. Aldhaifallah, M., & Nisar, K. S. (2013). Identification of auto-regressive exogenous models based on twin support vector machine regression. *Life Science Journal, 10*(4).
47. Gao, S. B., Ding, J., & Zhang, Y. J. (2011). License plate location algorithm based on twin SVM. *Advanced Materials Research* (vol. 271, pp. 118-124). Switzerland: Trans Tech Publication.

Bibliography

1. Fu, Y., & Huang, T. S. (2008). Image classification using correlation tensor analysis. *IEEE Transactions on Image Processing*, *17*(2), 226–234.
2. Liu, X., & Li, M. (2014). Integrated constraint based clustering algorithm for high dimensional data. *Neurocomputing*, *142*, 478–485.
3. QiMin, C., Qiao, G., Yongliang, W., & Xianghua, W. (2015). Text clustering using VSM with feature clusters. *Neural Computing and Applications*, *26*(4), 995–1003.
4. Tu, E., Cao, L., Yang, J., & Kasabov, N. (2014). A novel graph-based k-means for nonlinear manifold clustering and representative selection. *Neurocomputing*, *143*, 109–122.
5. Zhang, P., Xu, Y. T., & Zhao, Y. H. (2012). Training twin support vector regression via linear programming. *Neural Computing and Applications*, *21*(2), 399–407.

Jayadeva et al., *Twin Support Vector Machines*, Studies in Computational Intelligence 659, DOI 10.1007/978-3-319-46186-1

Index

Jayadeva et al., *Twin Support Vector Machines*, Studies in Computational Intelligence 659, DOI 10.1007/978-3-319-46186-1

Printed in the United States
By Bookmasters